HACKING
PLANET EARTH

HACKING PLANET EARTH

Technologies That Can
Counteract Climate Change
and Create a Better Future

THOMAS M. KOSTIGEN

A TARCHERPERIGEE BOOK

tarcherperigee

An imprint of Penguin Random House LLC
penguinrandomhouse.com

Illustrations by Joven Santos

Most TarcherPerigee books are available at special quantity discounts for bulk
purchase for sales promotions, premiums, fund-raising, and educational needs.
Special books or book excerpts also can be created to fit specific needs.
For details, write: SpecialMarkets@penguinrandomhouse.com.

Library of Congress Cataloging-in-Publication Data
Names: Kostigen, Thomas, author.
Title: Hacking planet Earth: technologies that can counteract climate change and
create a better future / Thomas Kostigen.
Description: New York: TarcherPerigee, an imprint of
Penguin Random House, LLC, 2020. | Includes bibliographical references and index.
Identifiers: LCCN 2019029958 (print) | LCCN 2019029959 (ebook) |
ISBN 9780593187548 (hardcover) | ISBN 9780525538363 (ebook)
Subjects: LCSH: Environmental geotechnology.
Classification: LCC TD171.9 .K67 2020 (print) | LCC TD171.9 (ebook) |
DDC 628—dc23
LC record available at https://lccn.loc.gov/2019029958
LC ebook record available at https://lccn.loc.gov/2019029959

International edition ISBN: 9780593189221
Paperback ISBN: 9780525538356

Printed in the United States of America
1st Printing

BOOK DESIGN BY KATY RIEGEL

For my family

Contents

HACKING PLANET EARTH

Why Can't We?

"Why can't we?" It's a question that nagged at me. Why can't we reset the course of nature, utilizing human innovation and advanced technologies? We have, after all, already bent nature out of shape and aimed it squarely against us. That turnabout largely began with the Industrial Revolution, when we started pluming massive amounts of carbon dioxide into the atmosphere that nature isn't able to soak up and store properly. Excess carbon means extra heat and global temperature rise—and more extreme weather. It means oceans expand and rise. It means more droughts and floods. It means mass casualties and entire populations relocating to more amenable geography.

Reducing carbon emissions, or carbon mitigation, hasn't worked; we continue to overpollute the atmosphere. And in September 2016, we went past the tipping point: Four hundred parts per million of carbon dioxide was calculated to be present in the atmosphere during the month when carbon should be circulating in the air at its lowest levels. Summer vegetation sucks more carbon out of the atmosphere than at any other time, leaving September with the least amount. But 2016 was different. The ceiling of four hundred parts per million became the floor—and we broke through it. That means the effects of global warming are likely irreversible without intervention. The result, if current rates continue, is temperature rise of 5.4°F by the middle of the century. The consequences of that are more bouts of extreme weather, higher sea level rise, massive migrations from low-lying areas, and endangered global food supplies. The heat might even completely destroy the Amazon rainforest, also known as the Earth's "lungs" because of all the carbon that it takes from the atmosphere and stores. If the Amazon goes, climate change effects will multiply exponentially.

Given this new reality, this hostile environment of the future, we have to opt for a more extreme approach to fighting climate change. We need silver bullets.

Geoengineering is defined as "the deliberate large-scale manipulation of an environmental process that affects the Earth's climate, in an attempt to counteract the effects of global warming." It's the fighting method we need if we are to stand a chance against nature's wrath.

Many environmentalists, including Al Gore, are opposed to the idea. He and others believe that if we intervene with the climate, we will be addressing the symptoms and not the causes of our human environmental plight; the movement to lower carbon emissions will be abandoned for expediency. They fear that people's preventative environmental actions, such as using less fossil fuel energy, will wane and society will instead count on unproven cures. But the National Academy of Sciences says exploring and funding research for geoengineering possibilities are necessary.

Visionaries such as Sir Richard Branson, Elon Musk, and Bill Gates believe artificial modification has its benefits if executed carefully and responsibly. I'm in their camp. Throughout this book various geoengineering methods are explored, as are ways to better manage the other natural elements we rely on for our survival—land, seas, our freshwater and food sources, and more.

Over the years, I supported the idea that if masses of people took little steps to save the planet, we could do it. That idea was manifest in my 2007 book, *The Green Book: The Everyday Guide to Saving the Planet*

One Simple Step at a Time. It gave hundreds of simple actions people could take to reduce waste, lower their energy use, and save water.

A year later, in my book *You Are Here: Exposing the Vital Link Between What We Do and What That Does to Our Planet*, I explored how we affect people, places, and things all over the world by what we do when we waste, pollute, and overconsume. The book called for more environmental education and awareness, and showed how we are all connected by our mutual actions. But a few years later, I realized that what we were doing as environmental activists wasn't working. I then wrote a comprehensive book on climate preparedness—a National Geographic guide for the inevitable effects of an extreme environment.

September 2016 was another turning point for me. If we could no longer slow global warming, if it wasn't enough to prepare, we'd have to take control of nature.

Hacking Planet Earth shows us the world we are in for and what we can do about it with forward-thinking fixes. Fate cannot be left to chance any longer. We have to forge our own way ahead using the one faculty that separates us most from all other living things on Earth: our ability to think, to innovate—our ability to reason. It's only our ability to reason with an increasingly hostile planet—made hostile by our own doing—that will allow us our future.

But this movement won't work at the grassroots level. It is time to turn our collective attention toward supporting industry and encouraging the business community, scientists, and technologists—innovators!—to step up and do what they do best: invent, pioneer, disrupt the same old ways of doing things. Yes, industry, the sector of society responsible for much of human-caused global warming to begin with, has to turn things around and lead the charge to help mend our climate.

For decades, environmentalists have sounded the alarm about the effects of climate change. The plan was to foster demand by the public at large that would compel businesses to change their practices and, in turn, bring about socially positive climate policies and regulations. But that is a pinball game of a plan. It is based on environmental education that incites public action that in turn presses businesses to change and governments to adopt better policies. That plan has failed. We can still press the buttons and ring the bells, but we'll never get the climate into livable shape before irreversible catastrophe. The United Nations Inter-governmental Panel on Climate Change says we have only until 2030 to act before unprecedented changes in all aspects of society will occur. Visionaries have changed society quickly in previous eras. Climate change necessitates such expediency—now.

Henry Ford didn't take surveys of horse owners and get consensus buy-in before he built automobiles. If he had tried, people would have thought him mad. Imagine him saying this: "It's a machine, you see, with wheels. It will cost you a fortune, and by the way, you will need petroleum to run it. So, we'll have to drill for that. And then we have to build roads on which to drive. The auto won't go faster than the horse you are riding. And to make autos in mass quantity, we'll have to build huge factories." Nuts. A complete shift in society, urban planning, natural resource extraction, labor force, and infrastructure. But he did it. Bill Gates cut a similar path with the personal computer. Telephony has put phones in everyone's hands. The Internet has connected us. Innovations and individuals forced the world to change—fast.

We need a radical disruption by engineers, investors, and visionaries to break us through the climate hold we are in. We need immediate solutions—far-out, far-thinking, world-changing solutions.

In the following pages, you will meet these solution providers. You will meet scientists and entrepreneurs. You'll meet adventurers and activists. You will travel the world with me, from the Arctic Circle to the Sahara Desert; from the bayous in the Deep South of the United States to underground research laboratories in Switzerland. I pull back the curtain on climate solutions that can save us from an overly polluted and dying world. These are the answers some people fear. These are the so-called moral hazards that it is said will dissuade people from taking steps to mitigate their polluting and abusing ways. I sincerely believe that casting these methods aside and keeping these fixes in the dark is doing the world a great disservice. And that is why I wrote this book. I want to expose what is possible.

I think we can do both: lower our environmental footprints *and* invest in radical solutions. This is the secret to a reengineered world. We should be celebrating and supporting these innovations and innovators, not fearing or ignoring them. They are what and who will save us.

In Mary Shelley's *Frankenstein*, it was the villagers who were the evildoers, not the monster. We already live on a FrankenPlanet. We have the opportunity to make it better. There really is no other choice for human survival. There is no going back.

Onward we must go . . .

CHAPTER 1

Thor's Hammer

It's home to the world's worst weather: winds that eclipse two hundred miles per hour, enough to strip the bark off trees; temperatures that regularly dip below freezing; blizzards; whiteouts; fog; avalanches; and ice that creeps onto almost every hard surface in sight.

You may think what is being described is a remote place in Antarctica, or somewhere near its opposite, the North Pole. But the world's worst weather can be found just a few hours' drive north of Boston, at the top of Mount Washington in New Hampshire.

Native Americans first discovered the mountain's wickedness, calling it *Agiocochook*, meaning "home of the great spirit." America's first weather station was built here. And until recently, the fastest wind speed ever recorded on Earth—231 miles per hour—was clocked at the summit observatory.

It's the curious position of the mountain and its features that breed weather's perilous progeny here. Mount Washington is the highest peak, at 6,288 feet, in the Northeast. As the tallest barrier east of the Mississippi, it blocks westerly winds. Its coastal proximity, less than a hundred miles from shore, makes it favorable to low pressure zones. And perhaps most causal, the range it inhabits is the vortex of storm tracks from the Atlantic, Gulf Coast, and Pacific Northwest.

A Laser Lightning Rod

Even on sunny days, afternoons quickly turn bitter on the mountain slope as the steep western face blocks the sun, casting shadows that dim among the rocks and ice. Snow races away. High winds don't allow crystals to pack and settle much.

When darkness appears on the face of the cragged cliffs above, it is an ominous reminder of what lurks: the great spirit—or what have you—evil weather.

The world's most abominable weather, however, is coming down from the mountain. It's branching out, pulverizing coasts, flooding plains, tearing through towns, and leaving little but scorched earth behind. We see the tragic images on the news almost every day.

Over the past decade or so, some of the strongest and most deadly storms and weather events have plagued the planet. Tropical cyclones such as Hurricane Irma broke speed records in 2017, topping the most destructive storm season ever recorded.

Winter weather has been no kinder. Snowfall accumulation, temperature drops, and ice storms have also set world records of late. In 2018, Moscow—Moscow!—saw more snow than it ever has. And that's saying something.

The Northeast and mid-Atlantic United States have begun coining terms for their worsening annual blizzard plights: snowmageddon, snowpocalypse, snowzilla. The curious list of monikers goes on.

Epic five-hundred-year floods are occurring annually. And heat waves and cold snaps are affecting people like never before. All-time hottest temperature records were set all over the world during just one week in July 2018.

Global temperatures have been rising on average for decades. Ultracold temperatures, too, have not been lost to global warming: A colossal cold front froze much of the United States, bringing Arctic chill as far south as Georgia, in the winter of 2017–18. Just a couple of years before that, Miami and Key West, Florida, put out freeze warnings.

Weather weirdness is becoming the new normal.

Extreme weather has haunted the planet throughout history, but this breed is different; it's born from humans. It's human-caused climate change. Climate change has brought with regularity weather that once existed only in aberration—in places such as Mount Washington—to mass centers of population the world over.

Global warming is arming weather with more moisture whisked from the ground, which is then unleashed from the air as heavier bouts of rain and snow. Warmer temperatures clash more fiercely with colder temperatures, producing more intense tornadoes, hurricanes, and tropical cyclones of all sorts and variants. Storms are, without question, stronger on average than in the past. From thirty years ago, average wind speeds and precipitation have increased 5 percent. The increase has made natural disasters more destructive and seasonal changes more pronounced. Oceans can't even escape temperature rise in the atmosphere. Warmer seas are swelling, and sea level rise is fostering coastal surges and successive superstorms.

It's frightening to conjure these images, but scientists have sketched various portraits of the world if global temperature rise continues and remains unchecked: a three-hundred-mile lake formed by epic floods in California; the Great Plains barren; Miami, submerged. Conditions like those atop Mount Washington will be even more widespread. And heat, too, will take its toll: New York City will be a crowded hotbed of disease and hyperthermic death. A very dystopian future can be conjured.

But half a world away in Switzerland, in the basement of a grungy building, almost hidden in a dark corner, lies a solution for another reality, another future. The ultimate weapon of defense to weather's wrath is a thin red line of light.

"It's a high-powered laser," explains Jean-Pierre Wolf, pointing to it. Middle-aged and of average height and build, Wolf doesn't appear to be a modern-day Thor, the Norse god who could control lightning with his hammer. Nor does the Swiss physicist, who was born in France, come across as some lab geek. He is athletically built, and the boots, corduroys, and ski sweater he wears make him look more ski instructor than science professor. This is surprising because Wolf has spent his career in research settings, earning his PhD in physics at one of Europe's most famous science and technology institutions, L'Ecole Polytechnique Fédérale de Lausanne, and later teaching at the likes of Yale, and universities in France and Germany. Like any scientist, his studies exposed him to different fields of inquiry, spectroscopy being one. The field of examining the interaction of matter and electromagnetic radiation is used in medicine, machinery, circuitry, and . . . weather. A real-life experience, not theory, made manipulating weather his mission. If Wolf's laser invention lives up to expectations, he might really be able to stop wicked weather in its tracks and create a more hospitable future for us all.

Until now, sordid chemicals have been the go-to agents for trying to modify weather. Spraying clouds with silver iodide, for example, is a common practice for trying to make it rain. Dropped by plane or launched by rockets, the chemicals spread like buckshot amongst clouds and form ice crystals that become heavy enough to drop from the sky. Depending on the temperature below, rain, ice, snow, or hail may fall.

The problem is cloud seeding doesn't always work, and when it does, the results aren't so predictable. In the 1950s, a secret British military cloud-seeding operation called Cumulus drenched the English countryside, causing flash floods that killed thirty-five people. The United States tried the same airborne tactic over the Ho Chi Minh Trail during the Vietnam War, dubbing its mission Project Popeye, in hopes of extending the monsoon season and causing floods and mudslides for the enemy. The mission's unofficial slogan was "make mud, not war." But Popeye's success was uneven and eventually abandoned. More recently, when China hosted the 2008 Olympics, it said it "cleared the skies" by firing more than a thousand rockets containing cloud-seeding chemicals into the air around Beijing to make it rain prematurely. Despite the pleasant conditions for the opening ceremonies, the weather fix was never scientifically proven accurate.

Wolf's laser is an entirely different technique for changing the weather. And it works systematically, brilliantly in fact, in both laboratory tests and outside in the real world.

He invented a laser that is more powerful than all the nuclear reactors on Earth combined. It can manufacture lightning in a cloud, break through air molecules and make rain, or conversely, blast through water molecules and dissipate precipitation.

The total apparatus is the size of a foosball table, and it sits innocuously in a basement at the University of Geneva, where Wolf teaches physics.

He didn't set out to invent a device that could manufacture or control weather. Wolf's doctoral thesis was in laser technology, but lasers can be applied to all sorts of things, from the remote control that allows you to change the channel on your television to medical applications, where lasers can identify and kill cancer cells. A flight from Geneva to Rome in the year 2000 honed the path of his research.

Wolf's plane got caught in a severe thunderstorm and was struck by lightning, and as a result, he began thinking about how laser technology could be used to throw lightning off its course. Lasers and lightning have a lot in common, he knew. They both direct energy. What if a laser's energy could be used to push lightning in a new direction? he wondered. It wasn't, of course, the first time someone looked to lightning for answers.

In 1752, Benjamin Franklin famously flew a kite in the sky during a thunderstorm to learn about nature's ability to harness electricity. His experiment led to more research and eventually to the way we produce and transmit electricity today.

Wolf, however, wasn't simply looking to capture lightning. He was looking to control it. That would take some doing because to control lightning, Wolf quickly realized, he had to re-create lightning. And that isn't so easy. A single bolt of lightning is hotter than the surface of the sun. The thundercloud from which it shoots can pack the energy of a hundred atomic bombs. To match that, he'd need an insane amount of power that could stretch from the ground to clouds miles high in the sky. If it wasn't for a breakthrough that lasted a billionth of a second, Wolf's laser would never have reached the limits required to control nature. The secret to Wolf's success is a fast pulse.

Usain Bolt can run faster than any other human on the planet—for ten seconds. He can't run a marathon going that fast. But in bursts, he can make up distance pretty quickly. Wolf applied that same burst concept to lasers to get the force he needed to match lightning's scale. His laser manufactures super powerful energy in one-billionth-of-a-second bursts. For perspective, a camera's shutter speed measures in the thousandths of a second. The reason Wolf concocted these superquick bursts is so the laser's power can stretch far into the sky, punching its way higher and higher until it reaches clouds with the same amount of force as its first pulse. The trouble with most laser beams is that they weaken over distance.

Wolf demonstrated in his lab how the laser works, pointing to the tiny mirrors the size of monocles on the table. With all of its instrument panels laid bare, the laser looks like the inside of the hard drive of a computer without its casing. He excitedly talks about the light filament that is created and the magic diamond that hypercharges the laser to its full force. Diamonds are superconductors of energy.

In the James Bond film *Diamonds Are Forever*, the evil character Blofeld uses diamonds to create a space-based laser weapon. The plot is absurd, of course, but the technology isn't. The crystal structure of diamonds concentrates light and makes it exponentially more powerful. Wolf uses a complex series of amplification to bring the might of the sun to his laser beam. On the table the laser beam travels through a maze of small mirrors before reaching a diamond. There, it suddenly changes color from red to blue. Blue is the laser's hottest phase, explains Wolf. That's when it's ready to blast out through its barrel.

The barrel is aimed at an enclosed chamber filled with liquid meant to replicate the atmosphere. The chamber looks like a small fish tank. When the laser blasts out and hits its target, the liquid inside the tank, globs of mist begin to appear, expanding and contracting. Microscopic water droplets scurry about, latching onto one another and breaking apart in a chaotic yet mesmerizing dance of nature re-created before your eyes. This is how a cloud forms. Soon the whole tank is filled with that mist. Puffy shapes of cotton-candy texture begin to break apart from the mold and come into their own. A swan, perhaps? An elephant? A wild horse leaving the pillows and plumes of cumulus behind?

A cloud being made by human hands may seem like a particularly awe-inspiring miracle, but it's being done here in a clinical laboratory, in a controlled setting. To test the laser in the real world, Wolf constructed a portable station the size of a freight container. He then took it to do battle with another Native American spirit on the top of another mountain—this one in New Mexico.

Wolf hiked to the top of South Baldy Peak, the state's highest, and fired away. "Artificial Lightning: Laser Triggers Electrical Activity in Thunderstorm for the First Time" was the headline that followed the next day in *ScienceDaily*. It went on to say that, "Lightning strikes have been the subject of scientific investigation dating back to the time of Benjamin Franklin, but despite this, remain not fully understood. Although scientists have been able to trigger lightning strikes since the 1970s by shooting small rockets into thunderclouds that spool long wires connected to the ground, typically only 50 percent of rocket launches actually trigger a lightning strike. The use of laser technology would make the process quicker, more efficient, and cost-effective, and would be expected to open a number of new applications."

To be sure, Wolf's experiment was a lauded success.

"Yeah, but our goal was this nice picture that you would put on your wall, where you see the lightning starting from the thundercloud and guided down to the earth," Wolf laments.

Instead, Wolf and his team had produced an intracloud bolt of lightning. He may not have gotten the snapshot he wanted, but he got the effect he desired and achieved something far more significant: the human ability to replicate in the atmosphere one of nature's most deadly weapons. That test was a decade ago, and since then, he has refined the technology even more.

The laser can redirect lightning that is already in the sky, pointing it away from dangerous strike points such as planes. It can redistribute air molecules within clouds in such a way that it produces precipitation. Or it can, with a different setting, break apart water molecules and keep things dry. In the future, he envisions mass production, developing laser lightning rods small enough so they can be attached to planes, trains, buildings, or virtually anything that is vulnerable to thunderstorms to deflect lightning, as well as other, portable laser stations that can burst or build clouds

So the future might look something like this: drones equipped with laser lightning rods roaming the skies and firing off beams like a squadron of R2-D2s gone mad. African deserts blooming with agriculture, and sopped cities like Seattle and London seeing brighter skies all year round. Living conditions, no matter where on Earth, could be made more temperate. All on account of Wolf's scientific breakthrough.

After his laboratory demonstration, Wolf turns off the laser. Its superpower gone, it is now just a faint red glow of light. When he flicks the room switch, and the lab goes dark, all that can be seen is that glow, like the eye of the Terminator, pulsing, waiting to go into action once again.

There are grand plans for what the laser can do. Wolf dreams big. He says that if we can change the weather, we can change the climate. Seasons, perhaps, won't carry the same meaning. And people trapped by the circumstances of their existence—the poor, the vulnerable, those living on the front lines of climate change who have historically lived with few natural resources—may be brought some of those resources and given the chance to prosper. Making it rain can provide freshwater to the billion or so people who live in desert regions around the world. Most of these people don't have enough readily available freshwater to drink, never mind for use in agriculture, for food. With several billion more people expected to inhabit the planet by century's end, natural resources will be further constrained, unless we can make the arid arable. Weather intervention by lasers can do that.

Take the megaflood tragedy in South Asia that affected forty-one million people in 2017. Homes, schools, and hospitals were destroyed. Roads, bridges, railways, and airports were severely damaged. Hundreds of thousands of people fled to refugee centers for food, clean water, shelter, and medical aid. Imagine if that could have been avoided, if lasers could have changed seasonal patterns. A particularly destructive monsoon season caused the tragedies. Monsoons happen annually and are shifts in prevailing wind directions. They typically cause heavy rains during summer in places such as India, or dry spells during winter. Lasers could break the cycle of rains that are brought about by wet monsoons.

Therein lies the power of weather modification . . . for good. But skeptics see only trouble if we start messing with the sky. James Fleming, a science professor at Colby College and author of *Fixing the Sky*, a book about climate control, says that in the past, weather modification has resulted in deadly storms and massive floods—the opposite of what might be hoped for by geoengineering. He cites Great Britain's Operation Cumulus, which was covered up for decades, as an example of what could go terribly wrong. And he isn't alone. Online groups and on-the-ground activists have decried tinkering with the elements. Even federal agencies have voiced concern about what might happen if nature's wrath is unleashed in unexpected ways.

The calls for concern harken back to paranoia about HAARP, the High-frequency Active Auroral Research Program run by the US military out of a remote facility in Alaska. HAARP's mission was to study the ionosphere—a layer of the atmosphere where radio waves can be controlled. For years, conspiracy theorists claimed the military was using HAARP as a cover to develop a secret weapon for weather modification and/or mind control. Nothing of the sort, it should go without saying, was ever proven. In any event, HAARP was shuttered in 2014. But alarm about weather modification hasn't quelled. Geoengineering Watch, an online group that claims more than thirty million page visits, regularly condemns weather and climate engineering with headlines like "Exposing the Global Climate Modification Assault" and "Waging Weather Warfare on World Populations." The group has events calendared months in advance and is warning about "biosphere destruction." Absurd claims, no doubt, will haunt weather change-makers well into the future.

A brisk breeze and snow begins to fall outside the laboratory at the University of Geneva. Wolf is hosting a lunch at a nearby restaurant for several of his colleagues and an observer. Besides the bread, wine, and pasta, on the table is the topic of weather and geoengineering. "Is it right to manipulate nature?" Someone notes that it has been going on since the agricultural revolution. "Are there unexpected downsides to firing lasers into clouds?" Nothing artificial is added to the atmosphere, someone else explains. The laser merely manipulates nature's existing weather recipes. On and on, questions and answers go. What becomes apparent is that it is up to innovators and interventionists to change the climate. Nature can't effectively do its job anymore; we've polluted and plundered the planet too much beyond expectation. Beyond even a creator's logic. Beyond even the gall of a great spirit.

Still, changing the weather might help the climate crisis, but it won't solve it. Weather is the symptom, not the cause of climate change. The battle of the future for a more amenable climate means winning the war on carbon emissions, solar radiation, and the many other infectious perpetrators of nature incited by human actions.

Thankfully, we as a human species are an inventive bunch, and all around the world there are pioneers like Wolf who are aiming to change the future for the better with technology of all sorts. They are working on geoengineering projects behind closed doors, deep underground, far out in deserts, in remote jungles, and even in outer space.

Brought into league, they can engineer a more benign environment for us all. The mission is finding them and learning how we can best hack the planet.

Sky River

 The Tibetan Plateau, sometimes called "the roof of the world" because it rises three miles high, may become Earth's largest weather testing station. China is strategically positioning thousands of burning chambers (about ten feet tall and a few feet wide) there that will release cloud-seeding chemicals to induce precipitation through a project promoters are calling Tianhe, or "Sky River" in English.

The Sky River rainmaking technology utilizes solid fuel—burning chambers that release silver iodide into the air, which when lofted by winds into clouds, excite particles that produce precipitation. A single chamber can reportedly produce a line of clouds three miles long. Tens of thousands of units are set to be dispatched.

Rocket science technology developed by China's military will allow the chambers to burn even at high altitudes, where oxygen is thin. Real-time data will be transmitted to a network of weather satellites that will also monitor extreme weather conditions.

The chambers only work when the wind is just right, blowing into the mountains and creating an updraft. The iodide chemicals hitch a ride on that wind current and rise high into clouds, or so it is hoped.

The Tianhe Project

Silver iodide has a similar molecular structure to ice—so similar, that ice already formed within a cloud is tricked and bonds to it. When enough particles form, they become heavy, and fall as precipitation to Earth.

The burning chambers that emit the rainmaking elixir aren't at all high-tech looking. They look like a tall chiminea—a bulbous outdoor stove with a tall cylinder vent on top. Each is estimated to cost about eight thousand dollars and can operate for years. A simple smartphone app ignites them, even at heights of more than sixteen thousand feet and in remote, often harsh, environments.

The Tibetan Plateau includes Mount Everest and the Himalayan Mountains. It stretches for sixteen hundred miles, from western China to India and beyond, and it is receiving less fresh water from glacial melt and fewer inches of snowfall each year. The plateau is already a dry region because cold air kept aloft at high altitudes holds far fewer water molecules than warmer air closer to the ground.

Freshwater scarcity jeopardizes the lives of millions of people—half the world's population lives in the region—and is giving rise to the radical weather modification technique.

The Sky River project was begun in 2016 as an experiment at China's Tsinghua University. The Chinese government is amping those tests so the project will become the world's largest weather modification program. If it works, Sky River will increase China's total annual rainfall by 7 percent. That's a massive amount, and enough to cover with rain an area three times the size of Spain.

The program will take years to implement and the side effects are unknown. Critics note that if you produce precipitation in one area, it takes away the propensity for rain or snow in another. That could be contentious in the Tibetan Plateau, where political tensions have historically run high between China and Tibet, and India and Pakistan.

But as things stand, the plateau, a sacred place for many, will see its landscape altered with new sights to behold in the form of weather chambers and maybe even a new sky above.

Hurricane Killers

Bill Gates and a group of scientists have developed technology to stop tropical cyclones in their tracks.

They have patented a device that destabilizes one of the main power sources for storms: warm surface water. By funneling warm surface water to cooler ocean depths and cycling the colder water back up, the process would be able to reduce the strength of tropical storms such as hurricanes and typhoons.

The device looks simple enough: a round tub floating atop two thin pipes that reach down into the sea. Waves fill the tub and the warm water is sucked down one pipe, powering turbines that connect to the other pipe—the cold water siphon.

The pumping system would lower the surrounding sea surface temperatures enough, it is hoped, to quiet a hurricane. Ocean waters must reach at least 80°F for a hurricane to potentially form. By measurably reducing water temperatures, storms would lack the energy they need to grow. Warm water is just one factor, however, in a hurricane's making.

"The actual process begins with a cluster of thunderstorms moving across the surface of the ocean. When the surface water is warm, the storm sucks up heat energy from the water, just like a straw sucks up a liquid. This creates moisture in the air. If wind conditions are right, the storm becomes a hurricane. This heat energy is the fuel for the storm. And the warmer the water, the more moisture is in the air. And that could mean bigger and stronger hurricanes," according to the Smithsonian Institution, which among other things, seeks to educate the public and educators on marine science.

Chilling warm seawaters is no guarantee that a storm won't form elsewhere, but it holds promise. A tropical storm's eye, or its core center, can be more than one hundred miles wide. That means the ocean-cooling devices would have a huge area to cover in order to be effective. Although marine scientists admit the challenges are big, they say the approach is theoretically possible.

Gates is involved with the project through his company Intellectual Ventures. The company's executives say the mitigation technology is meant as a backup solution in the event that climate change becomes uncontrollable, setting off wild storms and superhurricanes.

Hurricanes have become so intense and are projected to get so strong that some scientists and forecasters are calling for an additional strength rating to be added to the hurricane-measuring Saffir-Simpson wind scale: a category 6. Maximum category 5 hurricanes define storms with winds 157 miles per hour or faster. A 6 rating might cover wind speeds that reach 190 miles per hour, or 200 miles per hour, or more.

Ocean temperatures, just like air temperatures, are exacerbated by carbon emissions. If carbon reduction schemes continue to fail and ocean temperatures continue to rise, Gates's ocean cooling plan might be a welcome fix for future superstorms before they hit land.

CHAPTER 2

A Parasol for the Planet

The highest official temperature ever recorded on Earth was in Death Valley.

On July 10, 1913, a temperature of 134°F was recorded at the weather station in Furnace Creek, California. Furnace Creek is smack in the middle of Death Valley, the largest national park in the continental United States. It spans from California to Nevada, and encompasses the lowest point in North America, Badwater Basin.

Coming down from the mountains that soar 11,000 feet high into the valley and on through to the basin, 282 feet below sea level, is eerie. The valley floor seems like an optical illusion from up high: Is it a cloud? A lake? A glacier? It's only when the flat surface is revealed that the salt covering it can be discerned. The sun's reflection reads white, patches of white, to the mind's eye. The salt's reflection, along with the desert sand, is the reason why it gets so hot in the basin: It stores heat deep, kidnaps it.

Death Valley's extreme dryness means there are few clouds to get in the sun's way of sending its rays all the way to the Mojave Desert, where Furnace Creek lies. It is solar beaming unfiltered. That solar energy, or heat, is absorbed into the salt, rocks, sand, and soil. The rest is reflected, or reradiated, back up into the air. This vertical circulation is what adds to the heat effect.

The valley's shape exacerbates heat production. It's very low and narrow, surrounded by steep and tall mountains. When hot air rises, it doesn't have a chance to escape fully. The mountains trap the hot air before it dissipates. Air is sent back down to recirculate. This is how the air heats up even more.

Death Valley actually used to be a lake, Lake Manly, that existed thousands of years ago. It's still possible to read the water lines in boulders. The colors of rock faces are indicative of earlier, wetter times, too. Greens and blues are chlorides; pinks and purples are manganese oxides; and reds and oranges are iron oxides—chemistry changed by lake sediments as hot, mineral-rich waters bubbled up from below.

Now there is dry rock and hot sand and silver-like salt flats. There are properly sifted dunes that rise up high; tourists climbing them look like specks when they reach the tops. Mountain peaks lift in series, each range layered and lacquered and streaked in a powerful display of Earth's restlessness. Shifting tectonic plates crafted minerals into gargantuan rocks, rocks that became mountains millions of years ago in a geological process of faults and filling that try the imagination. Standing in the middle of the valley floor and looking up and around, with barely a soul about, only one thought comes to mind: This is Earth—pure, unadulterated, and awesome Earth.

Reds, rusts, and browns are so bountiful and complex that they are labeled and described by park officials as the Artist's Palette. Areas of sand are covered in dark soil, like layers of tiramisu. And there are endless stretches of white that bend and dust into the horizon—all that salt. Heat forms a sfumato for the naked eye, blurring the stark contrast of a mesa's lines against the cloudless blue sky.

You'd think they would at least put up a plaque, or something, in Furnace Creek, indicating where the highest temperature on this dear Earth was recorded. But there isn't one. A visitor center now takes up the exact spot on the Greenland Ranch where the 1913 weather station used to be.

A more modern weather station with newer technology is now located out back. That station in 2013 recorded a temperature of 129.2°F. The reading ties with later ones in Mitribah, Kuwait, in 2016, and Ahvaz, Iran, in 2017. Because these readings utilize modern technology, many say this is the legitimate high number to beat. Modern weather stations utilize more exact and in some cases digitally enhanced thermometers, rain gauges, wind vanes, anemometers, barometers, and hygrometers.

Satellite readings have rendered other records. There was the one in the Flaming Mountains of China in 2008 (152.2°F). The one in the Lut Desert of Iran in 2005 (159.3°F). Queensland, Australia in 2003 (156.7°F). But the World Meteorological Organization (WMO), the official title keeper of temperature records, doesn't recognize satellite measurements, only readings from thermometers physically on location. By that technical assessment, none has ever topped the 1913 reading in Furnace Creek.

"Weather historians have questioned the accuracy of colonial temperature records from many weather stations versus the modern records at these same sites," the WMO admits. But as of this writing, it was sticking to its guns with regard to Furnace Creek's top hot "honor."

What isn't in dispute is the fact that heat around the world is rising. Eighteen of the warmest years on record have occurred since 2001. Wicked heat waves have scorched North America, Europe, Asia, and Australia. A 2015 heat wave in India killed more than two thousand people. Fifteen hundred people were hospitalized in Japan in July 2018 due to heat illness. Six died. That same month, an oppressive heat wave shattered records in North America. Eighty million people were issued heat advisories and warnings. In Canada, dozens of people died, along with several in the northeast United States.

A heat wave is trapped air. Rather than moving about as wind, it stays put—and gets baked by the sun's energy. High pressure systems force air toward the ground, blocking it from circulating back upward, much like what happens in Death Valley. Technically, a heat wave is defined as two days and nights of high temperatures brought about by these meteorological phenomena.

As more of the sun's energy gets locked in the atmosphere by greenhouse gases such as carbon dioxide, the more temperatures rise and the more likely heat waves become. Because greenhouse gas emissions aren't being curtailed fast enough, heat waves are expected to lengthen and become more frequent. Heat illness, or hyperthermia, and dehydration are closely linked with heat waves. Such health concerns are among many reasons why major cities are exploring radical solutions.

Reflective materials are being used in infrastructure to cool local temperatures. Trees are being planted for shading. And public service announcements about heat illness are promoted more and more. "If humanity's greenhouse gas emissions continue to increase, the average temperature of the Earth's lower atmosphere could rise more than 4°C (7.2°F) by the end of the 21st century," the United Nations says.

Greenhouse gases may soak up more of the sun's energy than in the past, keeping heat trapped. But the main force behind the heating of the planet is, of course, the sun itself.

The sun constantly blasts energy toward Earth. Those energy waves become heat. But what if we could stop the sun's energy before it hit the planet? What if, as the inimitable music group the Bee Gees asked, we could stop the sun from shining? There are plans for that. Fantastical, mind-bending, seemingly far-fetched, straight-out-of-science-fiction plans. Really.

Imagine a space parasol that could deflect the sun's rays before they even reach the Earth's surface—a mirrored shade that would accompany the Earth's rotation around the sun. A sunshade could control the amount of heat on the planet without so much consideration for the effects of greenhouse gases. It could cool the planet immediately and send the burden of global warming into outer space. The shade would be controlled by us here on Earth and would autonomously operate in space much like a satellite.

There have been many plans over decades for all sorts of sunshades: a huge glass shield; clouds of moon dust; tiny umbrellas; miniature spacecraft. All of these blocking mechanisms would be placed at the Lagrange point—the gravitational neutral point in space between the Earth and the sun. Objects there don't move, as such. They are kept in gravitational place. A sunshade would bounce solar radiation away from Earth and into the dark reaches of space. The amount of energy the Earth received would be diminished, and atmospheric temperatures would cool.

For a time, mostly during the 1990s and into the first decade of the twenty-first century, these space projects were ambitiously undertaken. That was when the green movement began to take hold globally, Al Gore preached his sermon that was *An Inconvenient Truth*, and going carbon neutral became a fad. A radical shift in environmental thinking had occurred and so too had ecological solutions. The zeitgeist seeped into academia, of course, and great minds began thinking about interesting solutions to global warming. Dr. Roger Angel, a professor of astronomy and optical sciences at the University of Arizona, who has been referred to as one of the world's most brilliant and audacious engineers, began wrapping his brain around the warming conundrum. In 2005, he says, he began working on a new approach, something way out there, literally and figuratively.

He teamed with Dr. Simon "Pete" Worden, another astronomer, to develop the concept for a giant space parasol one thousand miles across. It would be built using lunar resources and wouldn't be a single structure. It would be a constellation of free-flying parasols of "gossamer-thin, lunar-made glass."

The parasols would be made in space, in free-orbiting factories near the Lagrange point. Glass would be delivered from the moon. "We envision the constellation as being like a large shoal of fish or flock of birds, with station-keeping control largely by autonomous computers in each unit to prevent collisions or self-shadowing. A local positioning system like GPS would also be used," they explained.

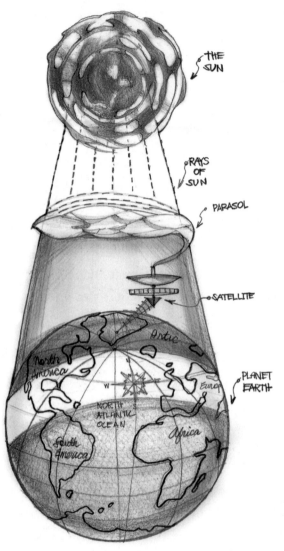

THE SUN

RAYS OF SUN

PARASOL

SATELLITE

Artic

North America

W

Europ

NORTH ATLANTIC OCEAN

South America

Africa

PLANET EARTH

The Space Shade

To make the ten billion units they'd need to comprise the constellation would require deploying a million of them a day for thirty years.

"The shield would require three major high-tech elements that would likely be manufactured and launched from the Earth. The first would be the package to enable material production and launch on the Moon. This would include the robots, electronics, solar cells, wire, bearings, motors and high-temperature ceramics for the lunar manufacturing and for the rail gun to launch the manufactured items back off the Moon. It would also include the pilot facilities on the Moon to bootstrap the local manufacture of structural elements used in full-scale lunar operations." They detailed all of this in manual-like form to explain how exactly the project would be accomplished. Their paper was published in *Ad Astra,* the National Space Society's journal, in the summer of 2006.

Eventually, Angel says, he realized the project had some practical challenges, so he developed another iteration on his own. This would be a "sunshade cloud" of tiny spacecraft. The meter-sized "flyers" would be assembled completely before launch, avoiding any need for construction or unfolding in space, and they would weigh about a gram each. "It seems feasible that it could be developed and deployed in [less than] 25 years at a cost of a few trillion dollars," Angel says.

More than a decade after his thesis was published, Angel is still sanguine about the possibility for the engineering feat. "The electromagnetic launcher is physically possible, but still well beyond anything now operational," he says. "The individual one-gram spacecraft are closer to reality than they were in 2006!"

Although the sunshade plans have never come to fruition, they have received serious attention within the science and academic communities. The findings have been published or written about in numerous journals. One of these articles caught the eye of Dan Lunt, a professor of climate science at the University of Bristol in England. He decided in 2007 to have a go at developing climate models around just what a sunshade world might look like.

Lunt's specialty is creating climate models to test hypotheses. (He also creates mean *Game of Thrones* GIFs to better explain science happenings. But those are stories for another time.) In any case, Lunt explains that he and his colleagues got to talking about the article by Angel and Worden and mirrors in space and the effects of sun-blocking on the Earth's climate.

"I was at a stage in my career where I had time to be able to follow leads like that, on a whim, if you like. And I just literally, that day, set up the model simulations," he says from his university office, where he is currently immersed in the topic of deep time and how climates acted (and reacted) millions of years ago.

His sunshade climate model revealed fascinating if not concerning results. Without a doubt the sunshade works to lower global temperatures, but not equally and not without some grave ramifications for life as we know it on Earth.

"We didn't know what to expect," he says. He carried out three different simulations under different cooling scenarios, reducing in each case how much solar radiation reaches Earth. The average amount of solar radiation the Earth receives is called the solar constant.

The sun sends a massive amount of radiation to Earth, enough in one hour, for instance, to power all the world's energy needs for a full year. Lunt discovered a reduction of 3.6 percent was needed to offset a fourfold increase in global carbon dioxide emissions, which is about how much of an increase in CO_2 there has been over the past two hundred years or so and is being attributed to human-caused global warming.

By dialing back the clock in simulation form, he was able to produce a climate that was exactly the same—well, in terms of air temperature anyway—as in preindustrial times.

Pre–Industrial Revolution carbon quantities in the atmosphere are what environmental activists are constantly striving for. Al Gore famously displayed the levels of carbon that have been released into the atmosphere over the last 650,000 years with a "hockey stick" graphic. Prior to the Industrial Revolution, some two hundred years ago, carbon dioxide levels were relatively consistent, fluctuating in a band between 180 parts per million and 290 parts per million. In the twentieth century, with the advent of machines emitting pollution, carbon levels shot up—looking in graph form like a hockey stick, the constant measures suddenly spiking high to more than 400 parts per million and growing.

Lunt discovered that if a sunshade were to be deployed—an engineering job of epic proportions, costing trillions of dollars and taking decades—it would measurably cool the oceans around the tropics. So far so good. Cooler tropical seas are needed to disperse cooler temperatures around the globe. The equator, which obviously runs through the tropics, is where the sun beats down forcefully, and a sunshade would reduce that energy. But Lunt also found that solar energy deflected by a sunshade would be disproportionately distributed around the world. This, in computer model form, caused anomalies: Ocean patterns were disturbed, polar temperatures were dramatically affected, and droughts appeared in more places.

Temperatures increased most from retarded solar energy distribution caused by the sunshade in the Arctic area, whereas off the south Atlantic coast of Africa, they decreased most. The natural transportation of the sun's energy from the equator to the respective poles was kicked out of balance, he found.

It should be noted that solar radiation is never uniform, sunshade or no sunshade. Amounts of solar energy depend on the angle of the sun. Because the equator is pretty much perpendicular to the sun, it receives the most heat, relatively, on the planet. Because polar regions are positioned at steep angles from the sun, they receive less heat. When heat arrives at the equator, it disperses, simplistically speaking, north and south, gradually cooling as it reaches areas farther and farther away. Once it reaches the poles, air circles back toward the equator again, and the convection cycle continues.

Lunt compared two scenarios: a preindustrial world and a geoengineered world. "We also find important differences in the hydrological cycle, with [a] sunshade world generally drier than the preindustrial," he observed.

The general effect of the sunshade actually does work to cool things, Lunt concluded, but the anomalies reach far and wide and portend great unknowns. Thus, he says, geoengineering in space is "a crazy idea." Beyond the theoretical application of a sunshade, which, as pointed out, could disturb ocean patterns, polar temperatures, and create droughts, the practical application of building the sunshade is considerably daunting. And the cost isn't exactly cheap: three trillion dollars for the one proposed by Angel and Worden, for example.

Despite the costs and complications, big plans are being made for manufacturing in space. Whether this will result in a sunshade is still a guess.

The International Space Station already has a small manufacturing facility and uses 3-D printing methods to make equipment. And Made In Space, a California company, is exploring new ways of designing materials using "the unique traits of the space environment" (think weightless) to create entirely new structures. "Out of this world" may be an understatement.

No matter the manufacturing advances in space, Lunt remains highly skeptical of geoengineering our future. "Renewable energy and decarbonization is becoming cheap enough that I think that is where the future lies. So, personally, I think that that is where the research should be going and where the funding should be going," he says.

Angel has also moved on from his much ballyhooed sunshade ambitions. "I concluded my time would be better spent on applying optics and physics to making solar power less expensive than fossil fuel. This is what I have been doing ever since. Better to stop emitting CO_2 than shading its warming effect," Angel says.

It's the sunshade's tag-along effects that worry Lunt most. They include abnormal migration patterns, pollination cycles, and strange seasonalities that could alter ecosystems in untold ways. Agriculture might be threatened and, in turn, food sources. Solar engineering is not a light switch. "You just can't turn it off," Lunt says.

Other scientists have examined what might happen if the Earth's temperature were suddenly changed, and they discovered that mass extinctions of species may occur. Plants and animals, all organisms, are hypersensitive to their environments. Change the climate even a little and cataclysmic events proceed. Global warming over the past century, for instance, has helped incite what some are calling the sixth extinction. Five extinction events that we know of have plagued the world, purging life-forms. And we are now losing species at a rate not seen since the time of dinosaurs.

There are approximately nine million different types of species on Earth. These can be tiny marine organisms or us human beings. Species types go extinct all the time. This is what scientists call the background rate. Typically, it means a falloff of as many as five species per year. We are now losing dozens of species every day, as much as ten thousand times the historical background rate. If this continues, there are projections that half of all species on the planet may be gone by mid-century.

Some species have had time to adapt to our warmer planet. But reverse course all of a sudden by solar engineering, and those species are doomed. There would be a rapid die-off.

A deep-time experiment that Lunt is working on may be most indicative of what could happen were the sun's energy to be diminished through solar engineering. "It is interesting to note that the combination of reduced solar forcing and high CO_2 has been present before, in the geological past," he says. This was during the Cambrian period, five hundred million years ago. "Therefore, geoengineering a future climate—Sunshade World . . . could be likened to turning the clock back to the Cambrian World," he says.

Global temperatures at that time were relatively mild, according to the paleo records. In fact, North America was rather tropical. It was during the Cambrian period when most of the animals on Earth first appeared. The Cambrian Explosion refers to the sheer number of life-forms that sprang up during that time. But all was not so hunky-dory. Another ice age followed that period, which is described by scholars as the third-largest extinction event in Earth's history. Colloquially put, the table is set for that eventuality.

Until a space test is conducted, every possible outcome of solar management is theoretical. The current emphasis on space travel, however, may change things from the theoretical to the actual. The first car in space is now, after all, floating around out there somewhere, thanks to Elon Musk. A sunshade, all of a sudden, doesn't seem that wacky. What's seriously unthinkable is the possibility for myriad losses of life-forms, perhaps including our own.

Moving Earth

An Earth closer to the sun would send temperatures soaring. An Earth too far away from the sun would have us freeze. Where we are positioned in the solar system has, until relatively recently, suited modern life on Earth just fine. But a small nudge in the right direction could cool the planet enough to make up for global warming.

The BBC reported about one idea: to explode hydrogen bombs in space and jolt us far enough away from the sun to cool the planet. Careful analysis shows that wouldn't work; the Earth is too big and traveling too fast for any significant distance to be made by even a million nuclear bombs exploding in space.

Some very serious-minded scientists at NASA, the US space agency, however, have come up with a plan that could actually push our planet away from the sun, if need be. The concept is called gravity assist, and it is how we currently change the trajectory of satellites in space. It's also how we have managed to get *Voyager 2*, the space probe, to the outer reaches of the solar system to study Jupiter and Saturn.

Gravity assist works by leveraging the gravity of a planet or another large object in space. When a smaller object in space nears a planet, it gets caught up in the planet's gravity field. That force slings the smaller object on by. But something happens to the bigger mass, too: The resistance force moves it a little bit off its trajectory path. Tiny objects don't carry enough weight to materially change a planet's trajectory. But a big enough one could shift things.

Dr. Greg Laughlin at NASA's Ames Research Center in California has gone on record with a plan that he and his colleagues designed that could push Earth to a cooler place in the universe. He says it isn't all that complicated, either.

First, engineers would hijack a comet or asteroid and direct it close enough to Earth so that it sweeps past but sends some of its gravitational energy to us. The asteroid or comet would then be slung to the outer reaches of the solar system, where it would pick up more energy and essentially boomerang back toward Earth, repeating the process. After enough passes, Earth would move to a higher orbit away from the sun. Voilà! Cool Earth.

There are, no doubt, questions about how this might work. First, how does one hijack a comet or asteroid? (Comets and asteroids are similar, but made of different materials.) In 2014, the European Space Agency landed a drone on a comet, proving there is a way. They have plans for many more comet and asteroid landings in the future. So the technology exists to snag one of these objects floating in space. Next, how do you pilot an asteroid to a place of your liking? Strap a rocket to it. At least that is what Laughlin's plan calls for. Obviously, you would then direct the rocket toward Earth.

The danger of directing an asteroid close by our planet is if the calculations are wrong and the asteroid crashes into our atmosphere. In that event, the only thing left of our biosphere would likely be bacteria.

Even if gravity assist worked in moving the planet, there would be stellar consequences. The moon would be thrown out of whack. Without our moon, Earth would rotate faster, meaning days would be only eight hours long. Wind speeds would increase dramatically around the planet because they would move more in line with how the planet spins. One-hundred-mile-an-hour winds could occur daily. This would infect hurricanes and typhoons, turning them into megastorms. Trees would topple. Tides would slow. Chemicals necessary to producing life in oceans would be lost. Evolution over time would end. But for the time being, we'd stay cool.

Colonizing the Moon

 Roger Angel and Pete Worden weren't the first to consider space-based solar radiation deflection: a sunshade. In 1989, James Early, a scientist at the Lawrence Livermore National Laboratory in California, where far-ranging nuclear weapons are designed, theorized about a thin glass shield that could be built from lunar materials, such as moon rock. He also suggested using a location near the Lagrange point to position the sunshade.

Early provided possibilities for the shield's design, and estimated it should be two thousand kilometers wide, which is about half the width of the continental United States. It would be paper thin, but weigh about a hundred megatons. (One megaton equals one million tons.)

Construction of such a behemoth would be challenging, to say the least. Launching it from Earth, nearly impossible. Which is why Early suggested colonizing the moon in order to build it there.

The lens would be built in sections and shipped from the moon to the Lagrange point, or L1, as it is known. Designs were costly even thirty years ago: between one trillion and ten trillion dollars.

Modern efficiencies may have brought that price down some. But the feasibility is the same: It isn't.

The *MIT Technology Review* analyzed the Early project for today's environment and concluded that "such a mega-construction doesn't seem immediately feasible."

In the future it might be possible to undertake major construction projects in space. Indeed, the prospect for colonizing the moon is closer to reality. For decades, scientists have eyed and made plans for a research station on the moon. It has been too costly a consideration. Advancements in technology and space exploration have changed the economics. A recent blueprint estimates that we could build a research station on the moon by 2022 for just ten billion dollars. The US defense budget alone is more than seventy times that amount.

Construction in space is another matter. Even minor assembly work on the International Space Station is difficult and overly time-consuming. Cue Sandra Bullock and George Clooney floating out of control through space in the movie *Gravity*.

CHAPTER 3

Carbon Vampires

A vast reserve of molten carbon has been discovered under Wyoming, and stretches into Canada—a phosphorous, lava-burning, virtual hell of a cavern the size of Mexico—that if released, would pollute the air so heavily with carbon dioxide it would bring about the end of the global climate as we know it.

Carbon emissions, the primary driver of climate change, come from many sources. CO_2 is the chemical formula for a greenhouse gas, which is invisible to the naked eye. We humans, along with other animals, plants, soils, and such, emit carbon dioxide as a result of processing oxygen. Volcanoes and wildfires plume CO_2 from their ashes. Even oceans burp up vast amounts of CO_2 in their exchange with air.

But it is coal that is most closely associated with carbon emissions. In addition to oxygen, coal also contains hydrogen, sulfur, nitrogen, and sometimes other elements. Coal is made from plant matter that compresses over time—millions of years—into its solid, rock form. And it comes in different varieties: lignite, bituminous, anthracite, graphite, and specialty categories. For centuries it has been used as a fuel.

Molten carbon is something altogether different. It lies deep beneath the Earth's surface, near the upper mantle, where temperatures can top 1500°F. There, minerals melt. The group of minerals containing carbonate is expansive, and may contain as much as one hundred trillion metric tons of melted carbon—far more carbon than what was believed to exist on the whole of planet Earth. If released, that melted carbon would alter the course of the atmosphere and make our planet likely an impossible place on which to live. Thankfully, the lavalike carbon sea is buried more than two hundred miles beneath the surface. Scientists discovered it in 2017 using the world's most advanced seismic sensors.

Dr. Saswata Hier-Majumder, a geophysicist who teaches at the University of London, says he was working on examining, with sensors, magmatic melts some eighteen hundred miles underground in the Earth's core-mantle boundary when a colleague pointed to an anomaly closer to the surface. Magmatic melts are areas deep below the Earth's surface where magma—liquefied rock—exists. Most of the Earth's outer layers are solid, with patches of melts. Applying his expertise to the anomaly, Hier-Majumder was able to calculate how much melt was there, and in turn, how much carbon could be emitted from it. The amount is stunning: one hundred times the amount contained in all the world's proven coal reserves.

It doesn't mean all the carbon that was discovered is going to end up in the atmosphere anytime soon. It could take millions of years for the carbon to cycle up from that depth, the low-velocity layer, to the surface—that is, unless there was a volcanic eruption.

Yellowstone National Park in Wyoming sits atop a supervolcano, beneath which lies the sea of molten carbon. If the Yellowstone volcano were to erupt, as it did more than six hundred thousand years ago, it would drape the entire United States in ash, an occurrence known as a nuclear winter.

Even in its dormant state, the Yellowstone volcano releases forty-five thousand metric tons of carbon dioxide each day. That's as much as ten thousand passenger vehicles emit in a year. If just 1 percent of the recently discovered molten carbon escaped through volcanization, it would be the equivalent of burning 2.3 trillion barrels of oil, Hier-Majumder and the other scientists who worked on the discovery say. That, of course, isn't good. But Hier-Majumder walks that back some in a conversation. He doesn't believe there is real risk of imminent emission to the surface. "This is a deep carbon storage site," he says, "with an approximate residence time of one billion years." Still . . .

Recently, cracks appeared atop Yellowstone's grounds, including a one-hundred-foot fissure. People have taken it as a warning of a forthcoming eruption. And there is no harm in taking heed of the signs. But Hier-Majumder says there is just no reason to panic. Further tests will be done on the area, he assures. Given that his work is so abstruse—finding magmatic melt (also called partial melt) in the Earth's deep interior—it's easy to fall into the trap of extrapolating dangers from the examples he provides while putting things into layman's terms. He comes across a bit like Jeff Goldblum in *Jurassic Park*, trying to explain chaos theory. (It's really not that simple.)

For his experiment, Hier-Majumder relied on space-age technology: 820 seismometers spread throughout North America whose data were compared to a large number of earthquakes. Nothing like that had ever been done. It gave the Earth's deep interior what amounts to a CAT scan.

He says he is also working on a dense data set from Hawaii and southeast Asia, where more melting may be happening. "I hope to be able to generate a global map of melting," he says. The amount of molten carbon beneath the Earth's surface, then, is unknown but far bigger than was estimated.

What's for sure is the amount of CO_2 that is released into the atmosphere each year by us humans, which amounts to around forty billion tons. To keep temperature rise in check, that number has to be cut by about twenty-five percent—right now. The opposite is occurring: The world is on track to increase emissions by around 50 percent by 2030. The numbers vary based on different scenarios and greenhouse gas mixes, but the gist is that we are producing more carbon dioxide faster than nature can consume through the historic carbon cycle.

All living things are made of carbon. We live, breathe, die. Carbon goes into the ground and eventually makes its way to the surface and back into the atmosphere. That cycle is supposed to more or less balance out to provide a steady temperature range. Pollutants and increased emissions, never mind more people on the planet, retard that cycle. Which is why there is approximately 30 percent more carbon dioxide in the air than 150 years ago, and why global temperatures have risen nearly 3.6°F (2°C) over the same period.

All sorts of gnarly consequences begin to occur when temperatures rise more than that amount, as has been discussed in the previous chapters: heat deaths, storm intensities, sea level rise, and environmental anomalies abound. Salvation lies in not only cutting back on emissions but capturing and storing more carbon. It's tricky business.

Nature's biggest carbon-dioxide-capturing device is the ocean. There, approximately 30 to 50 percent of the carbon dioxide produced by humans is absorbed. The next biggest sponges are trees, which absorb as much as 25 percent of human-produced carbon dioxide. Soil and other matter lap up the rest. Carbon needs to be captured from the air and stored; otherwise it accumulates in the atmosphere, releasing energy. Temperatures, as we're experiencing, then rise abnormally.

It's impossible to make more seas or land to serve as carbon-capturing sinks, of course; we'd need a whole other earth for that. But with trees, more of them can be planted. In theory, the more trees planted, the more carbon that is absorbed. Yet even that process is slow. It takes the average tree about forty years to absorb just one ton of carbon from the atmosphere. Remember, tens of billions of tons of carbon are emitted by human activities alone each year. The amount of land that would be required for trees to make a significant dent in capturing carbon from the atmosphere would equate to trees being planted on a space the size of three Indias, according to some estimates. And at that, if all those trees suddenly sprung up, there would still be decades of lag time to bring carbon floating in the air back, as it were, to Earth.

That's why Klaus Lackner decided to make a different kind of tree.

Lackner, the director of the Center for Negative Carbon Emissions and a professor at the School of Sustainable Engineering and the Built Environment at Arizona State University, has come up with a better approach than nature's own carbon capturing system: He has created an artificial tree that can absorb as much as one ton of carbon dioxide per day. A forest of them could negate the entire amount of carbon emissions humans produce, and then some. For the moment, unfortunately, there is just one of Lackner's trees in the outdoors. It sits by its lonesome in the desert near Mesa, Arizona. Not much is around it but sand and scrub brush. Even all by itself, Lackner's tree processes carbon at an exponentially faster rate than one standing in a thick forest, never mind the amount it absorbs. Carbon dioxide is a cumulative gas, meaning it piles up over time before forming in a state that can be drawn back down out of the atmosphere. Lackner realized that no matter how good people got at reducing carbon dioxide emissions, there would still be that

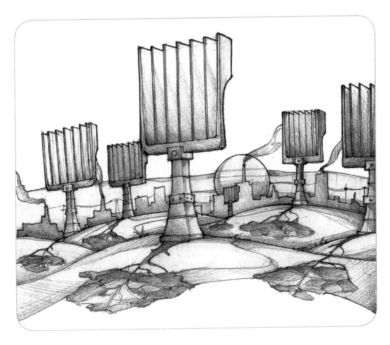

An Artificial Forest

decades-long lag. With the amount of carbon in the atmosphere already exceeding scientific limits to prevent destructive temperature rise, it became clear that a more urgent solution was needed. This brought about the idea for a huge fake forest.

Lackner is one of those mad geniuses who thinks in terms of math equations and physics calculations. A former scientist at the Los Alamos National Laboratory, famous for its role in inventing nuclear weapons, he one day in the early 1990s began contemplating a world without humans, a world where machines could replicate themselves. Could they? Indeed, it's possible, Lackner found. But they would need a sustainable source of energy so they could power the cloning of themselves. For all the talk of robots and artificial intelligence, energy creation apparently still needs a human touch. Or, at least a human kick start.

In considering energy production, Lackner also examined energy's detritus: carbon emissions. His curious mind took over, and he wondered what it would take to get the world's carbon budget under control. Lackner thinks in terms of budgets: supplies, costs, expenses. Carbon fits nicely into this equational analysis.

"The carbon budget has a limit," he says, leaning back in his swivel chair inside his stark fourth-floor ASU office, outside of which stand four natural trees. Irony, apparently, reaches high.

Lackner has the trace of a German accent and the demeanor of a late-middle-aged critical thinker—someone who buttresses statements with examples. "Because . . ." "That means . . ." The educator in him seeps through. To that end, Lackner explains the carbon budget's limit is in the range of 450 parts of carbon per million of air particles. He details why and how the limit is calculated. But the bottom line is that we are on track to quickly break that budget within seventeen years. A 450-part-per-million carbon budget means global temperature rise could be limited to the famous 3.6°F/2°C. Still a lot, but manageable. Exceed that budget and all of a sudden the climate as we know it will change—for the worse. Heat waves, high seas; all the scary scenarios that have been conjectured will come true. "I don't think we can convince the world to do something fast enough to save it," he laments. Which is the reason for his tree.

Lackner's tree wouldn't be easily missed in a forest. It's a metallic apparatus whose "leaf" is a carbon-capturing membrane. When air passes over the membrane, carbon dioxide is trapped and then funneled away and stored. Natural leaves operate in a similar manner. They just store carbon in stems, branches, and roots rather than a metallic tank. Lackner's tree sends carbon to a stainless-steel storage tank.

Trees and plants use sunlight to power the transformation of carbon dioxide in the atmosphere into molecules that they store and use to grow. Lackner uses water. When moisture in the air is captured and dries, it acts like a fuel and powers the storage process.

Lackner's tree, at a height of about twelve feet, looks like a small football goalpost. Between the uprights is the membrane, which is designed in an accordion-like fashion. When the membrane is stretched out as a virtual sail, it catches air and the carbon molecules in it. When it contracts, it squeezes out the carbon, funneling it into the tank.

There are various designs for how the tree could take shape—as a more natural-looking tree, such as a palm tree; as something that resembles a centipede; or even something that looks like SpongeBob SquarePants. This is what the miniature "tree" looks like inside his laboratory around the corner from his office. It's just a few feet tall.

The mini tree there, however, looks like SpongeBob is in the hospital: Various tubes, sensors, and monitors are attached to it to determine how much carbon is being captured and how much is released. The membrane rises and falls like a lung, sucking in air and pumping out CO_2.

The six-foot-by-four-foot version of the tree stacked as a forest could consume the entire thirty-six gigatons worth of carbon dioxide released into the atmosphere every year by fossil fuels. Granted, that forest would have to contain one hundred million trees. It may sound like a lot, but there are more than three trillion trees on planet Earth, so the possibility isn't all that far-fetched. Put another way, there are more than a billion automobiles in the world. Lackner makes that point especially meaningful because if there are enough facilities to produce all those vehicles, there is manufacturing capability to make lots of his devices. For example, the Port of Shanghai processes forty million shipping containers filled with stuff per year. "Which means there are factories nearby which can make enough equipment and goods that can fill those containers. We would only need to fill ten million containers a year with air capturing devices," he says. So the amount of air-capturing devices needed to chart the world on a more carbon-free course can be manufactured and shipped, no problem.

There are two issues standing in the way of mass production of Lackner's tree, though: political will and money. Each tree costs between twenty thousand and thirty thousand dollars to make. Individually, that is about the price of an average car. But as a group, the cost is enormous.

As for will: "We have difficulty acting when there are just probabilities and risks," he says. Scientific theory doesn't compel lawmakers to spend. Although it should. "Policymakers always have to deal with uncertainty. That is actually their job. Whenever we make decisions about interest rates, about taxes, about going to war, we are dealing with uncertainties," he says. Climate change is no different. It's just a risk we have never dealt with before. Hence, social action falls short in dealing with climate change.

Lackner equates excess carbon emissions to a waste management problem, to drive the point home. "With garbage and sewage, we have come to the conclusion that a) we can't entirely prevent it, and b) we are not allowed to just dump it. We have to properly dispose of it. And the same has to happen with CO_2," he says.

The rationale for removing even more CO_2 out of the atmosphere than we put in, tracks that reasoning, too; we have to clean up what has been dumped in years past. Which is why carbon capturing will stay expensive long into the future, because somebody is going to have to deal with and pay for the mess that has been made of the atmosphere. And likely, that somebody is the next generation.

Trees on the ASU campus are labeled. Small plaques are attached to their trunks with a tree's name and its benefits. One says, "Orangewood is derived from this tree. The sawdust from this wood is used to polish jewels. The fruits of this tree can be used to make marmalade." Another informs, "This is an extremely popular palm for landscaping in Arizona. It grows quickly with few problems. One drawback, however, is that it may get too tall for the average owner to trim." Students bustle by on foot, bikes, and skateboards without giving the trees or the plaques as much as a glance. Even the sign showcasing Lackner's tree "technology to take CO_2 back from air" that sits prominently in the center of campus, doesn't seem to get noticed.

A high-tech image of the device is shown. The poster it's on also shows natural trees, a smokestack in the distance, clouds, and patches of blue sky. A plane is pictured flying over a highway where futuristic-looking trucks and cars drive along. Of course, this is the idyllic version, an artist's rendering. A real tract of one hundred million trees would stretch for miles and would look like something out of a Mad Max movie.

According to Lackner's plans, artificial trees would not all exist in the same forest. They would be spread out around the world, ideally nearer to centers of high carbon-pollution concentration, such as chronically congested highways or coal plants. The closer to concentrated areas of carbon emissions, the more carbon that can be captured.

Picturing fake trees in a field as a surrogate for natural trees may bring about sci-fi imaginings—the original "could machines replicate themselves?" premise from which Lackner's trees come. But placing artificial trees alongside roadways and in industrial areas shouldn't seem so environmentally vulgar. A world with artificial trees might not look like a replanted green forest standing tall and stretching vast into the countryside, an image dear to supporters of reforestation. Yet, fake trees may have to be part of a comprehensive solution for redacting carbon from the atmosphere.

"We can do it," Lackner assures.

But let's say we do. Let's say fake trees start capturing carbon from the air. What do we do with all the carbon? Where do we store it? Lackner believes what makes the most sense is to turn the gas into fuel. Other schemes call for burying carbon underground, converting it to solid building materials, such as bricks, or pumping it deep into the ocean floor.

All those schemes are rife with issues. Greenpeace, the activist organization, says carbon capture and storage doesn't work. It flat out claims that the technology is unproven and too costly to be effective. Moreover, Greenpeace says storing carbon is dangerous business. "To actually deliver reductions, the emissions captured and injected must stay underground permanently. If leaked back into the atmosphere, they would only make climate change worse and threaten people and animals."

In a report, Greenpeace cites three examples of the dangers and downsides to capturing and storing carbon:

In Ain Salah, Algeria, in 2011, injecting carbon dioxide into sandstone caused earthquakes.

In Sleipner, at the Norwegian North Sea, one of the world's oldest injection sites for carbon dioxide, scientists found huge fractures in the seafloor, almost assuredly meaning the carbon dioxide would eventually leak.

In Mississippi, large amounts of carbon dioxide leaked from storage wells, killing deer and other animals in the vicinity.

It isn't just activist groups such as Greenpeace who have concerns about carbon capturing. A group of scientists has examined various carbon capture and storage technologies and has pretty much come to the same conclusion: It's costly and there are risks of leakage.

The reason direct air capture is costly is because of all the engineering, technology, equipment, and storage facilities that are necessary, as well as the energy costs of operating the conversion plant. The operation has to be done at a scale big enough to warrant servicing large demand. Large production of synthetic fuel is the only way to make it economically feasible and drive down prices. The cost of the fuel, of course, has to be competitive to market prices of gas at the pump, electric charging, or hydrogen filling.

The quagmire is that while the world needs to stop emitting more carbon into the atmosphere, it still has to deal with the amount that already exists. For that, it seems, there is no easy answer. But a tree grows in Mesa.

Capturing Carbon

 Climeworks is the opposite of a coal plant: It runs facilities that take carbon out of the atmosphere and convert it into fuel, rather than burning fuel and putting more carbon into the atmosphere.

"Our plants capture atmospheric carbon with a filter. Air is drawn into the plant and the CO_2 within the air is chemically bound to the filter. Once the filter is saturated with CO_2 it is heated (using mainly low-grade heat as an energy source) to around 100°C (212°F). The CO_2 is then released from the filter and collected as concentrated CO_2 gas to supply to customers," Climeworks explains.

Other companies have also developed direct air capture technology, converting carbon into fuel or fertilizer or for other uses. These technologies are considered "carbon neutral" schemes because the carbon that is captured is eventually released again.

The big breakthrough Climeworks has accomplished is in the area of negative emissions, where carbon is captured and stored and kept out of the atmosphere for thousands or millions of years. To permanently store carbon, the filtered concentrate is pumped more than two thousand feet underground. At its plant in Iceland, the carbon reacts with the basaltic bedrock, forming solid minerals. "A permanent, safe, and irreversible storage solution is created," Climeworks says.

Although promising, the effects of direct air capture technology aren't huge as of yet. The amount of carbon captured at Climeworks's Iceland plant, for example, is equivalent to the annual carbon emissions of just one US household. At its plant in Switzerland, where captured carbon dioxide is converted into gas and pumped into greenhouses, the equivalency is just twenty US households' annual emissions.

To be sure, as technologies sharpen and prices drop as demand warrants, plants will ramp up conversions. And as other companies step up and come online, captured supplies could equate to significant drawdowns.

Innovations in technology allow for carbon to be captured from the air much cheaper than in the past. Time was when it cost hundreds of dollars to convert carbon from the atmosphere into synthetic fuel. The price has come down and these days works out to about four dollars per gallon, which is in line with gas prices at the pump. Given those prices, direct air capture plants should start popping up in more places around the globe.

They don't take up too much space, either. Climeworks's CO_2 collectors are only 215 square feet in size per unit. Each unit, which looks like a commercial-grade air conditioner, filters about two pounds of carbon per day. In their own way, the units are actually air conditioners themselves—for the planet.

The Rocks of Oman

 They seem like normal rocks, but scientists have discovered that a certain type of rock in the country of Oman can capture billions of tons of carbon dioxide from the atmosphere.

These rocks are special because they are typically found far below the Earth's crust. Tectonic movements have pushed craggy slivers of the peridotite rock formations onto land and exposed them to the air. This brings about all sorts of mineral reactions. Scientists studying the peridotite discovered the potential for carbon mineralization—effectively turning carbon dioxide into stone—and have come up with various designs for using their carbon-storage potential.

The rocks are rich in olivine, a mineral that reacts with air and water to hypercarbonate.

Study of the mantle rocks is relatively new, but geologists at Columbia University's Lamont-Doherty Earth Observatory say the carbonation process can be accelerated by breaking the rocks apart and exposing more of the carbon-capturing minerals to air, or by drilling wells and pumping water through rock formations.

While Oman is home to vast amounts of peridotite, that type of rock can also be found in mines and aboveground in various places around the globe, including Northern California. The hope is that at least one location could serve as a test site to prove out the methodology.

"We propose that one method of speeding up olivine carbonation would be to 'jump-start' the process by drilling, fracturing, and heating a rock volume at depth to about 185°C, and then pumping purified CO_2 plus water into that rock volume. We calculate that such a process would convert billions of tons of CO_2 into solid carbonate minerals per cubic kilometer of rock per year," the geologists estimate.

That is a fantastic amount of carbon storage potential. We plume about forty billion tons of CO_2 into the atmosphere every year across the whole of Earth, and capture very little of that back through human ingenuity. Afforestation, or planting trees on crop, pasture, and brush lands, for example, sucks only about ten tons of carbon dioxide from the air per acre in a year—and that statistic is on the very high side.

Enough rocks could turn excess carbon dioxide in the atmosphere to stone, keeping it out of the air for thousands or millions of years into the future, depending on rates of erosion; a solid achievement.

CHAPTER 4

Desert Reflectors

 Six hundred million people live here without electricity, the most in one concentrated swath. They get by on sunlight, fire, perhaps even a portable generator or two.

Not being able to see at night is the least of their burdens. Health clinics and hospitals can't perform proper procedures after sunset. Childbirths in the dark, too, become risky. Birthrates fall. Refrigeration keeps medicines from spoiling—and food, also. But with no electricity, vaccines and food lose shelf life. Additionally, more primitive methods of cooking mean indoor air pollution soars. Burning biomass—such as wood, animal dung, compost matter, and the like—inside homes causes serious respiratory illnesses. Millions die prematurely each year from breathing cooking fumes alone. Of course, there is no Internet access or telephonic communication. Education, disaster warning systems, and community progress are all stilted without the benefit of electrical power.

Perversely, this massive segment of the world's population, more than half of the entire population of the African continent, lives just south of the place where the most amount of energy is naturally produced on Earth: the Sahara Desert.

Because of its size—the Sahara is about the same size as the United States—and the amount of pure, cloud-free sunshine it receives, enough energy is produced by the desert to service the entire world's needs more than eighty times over. But all this immense power is barely tapped. Solar technology hasn't yet captured the Sahara's full energy potential. Nor have transmission lines found their way south. Which means the sun is but a tantalizing prospect for those in the sub-Saharan African region who could use the Sahara's solar energy most. Other energy sources are too expensive or are out of reach. No one is contemplating building a coal-fired power plant or nuclear power plant here—yet. That's why so many millions of people are left in the dark. There is simply no access to power.

It gets so dark in the Sahara at nighttime that it's dangerous to walk about. There are no shadows to hint at what's to come. Placing one foot in front of the other becomes risky business. Who knows what might lie underfoot, or what may strike an appearance in front of you. A snake? A rock? A camel? Other senses take over. They perk up, long gestated and relegated to backup information providers. Your sense of hearing. Listen close, not for primary sound, but secondary sensation: a shift in wind, liters of sand being redistributed. Proprioception, or your sense of where your body is in space. A change in slope. An indescribable feeling that something, maybe someone, is there, yonder.

It's best to stay put in the desert at night and simply look up. Stars scatter in shiny pinheads. The dark background against which they are transfixed bleeds deep into infinity. It's futile to count them all, those stars. Far better to point out the shiniest of the bunch and try to remember their names. Sirius, the brightest in our universe. Or Polaris, the North Star. With time and some concentration, constellations become recognizable: Ursa Minor, the Little Dipper; Ursa Major, the Big Dipper; Orion.

In the Sahara Desert there are no city lights to brighten the sky. Night comes fast and swallows you in a canvas of faraway suns.

The sub-Saharan region includes forty-eight countries, by World Bank definition, that trace their way through the continent from Mauritania to South Africa. On average, at least 50 percent of the population in all but one of these countries does not have regular electricity. Across the globe, approximately a billion people are in similar circumstances and do not have connections to electrical power. Electricity is most scarce in developing countries, where about one in five of the people who live there can't plug in and power up.

Electricity is a Promethean gift to humanity. It enables participation in civilization. It distances us most from nature and allows us to live in artificial environments. Virtual reality aside, lounging at midnight on an easy chair, or watching television in a lamp-lit room and now and again traipsing to the kitchen to grab a snack out of the refrigerator, is tricking the natural order of night and day, hot and cold; energy is hood-winked.

Despite the areas yet to achieve full access to power (in Asia, two out of ten people don't have access to electricity, and in the Middle East and Latin America it is one out of ten), we humans have largely rigged the Earth with electricity and widely use fossil fuels as the source. Burning coal or oil or gas does the job of firing up steam to spin turbines and crank out electrical power. But the downside, as we know, is the refuse of those energy sources: toxins and temperature-popping pollutants that result in climate change. Two-thirds of the electrical power in the United States is derived from the burning of coal and other fossil-related fuels. Globally, it's about the same.

But make no mistake: The people with no power today are sure to get it in the future. And when they do, they could add even more to the mix of emissions plaguing us and polluting the planet.

The International Energy Agency forecasts that energy demands from developing countries coming online with electrical energy will skyrocket and account for 65 percent of the world's energy consumption by 2040. That compares with relatively flat demand from the United States and the rest of the developed world, where everyone is pretty much already plugged in. Moreover, population growth will be highest in the developing world, adding further to energy demands there.

Soon the entire world will be wired, reengineered to accommodate our needs 24/7, no matter the Earth's revolution around the sun.

It takes approximately eighteen terawatts of power to meet the world's energy needs. A single terawatt, for perspective, is equal to ten billion lightbulbs of the one-hundred-watt variety. By 2040, consumption will increase by nearly 30 percent. By the end of the century, energy needs are projected to soar four times that amount—up 124 percent. If fossil fuels continue as our major energy source, hope for a habitable planet like the one we are used to will be lost. There are other possible energy sources, of course. Renewable energies such as solar, wind, geothermal, and marine power exist. In total, they provide only about 1.5 percent of the world's energy supply. Nuclear power contributes a fraction of world supply, as well. It is very expensive to produce and comes with a woeful set of dangers.

In April 1986, the Chernobyl nuclear reactor failed and radiation was leaked into the local area and into the atmosphere, causing scores of deaths, injuries, cancer, and environmental destruction. It was one of the most horrific human-caused disasters in history, and much of the nuclear fallout still exists.

The incident disturbed forty-nine-year-old German particle physicist Gerhard Knies so much that he began investigating safer alternative forms of energy for which he could ply his trade. Dr. Knies made a calculation that would come to change his life: Within six hours the world's deserts receive more energy from the sun than humankind consumes in a year. That remarkable tidbit of information sent him on a mission to develop desert energy. It manifested into what he called Desertec, a big idea to cover with solar panels every inch of desert area on the planet.

Deserts are defined as areas receiving less than ten inches of precipitation per year. They come in different forms: hyperarid, arid, semiarid, and dry subhumid. All told, they make up more than 40 percent of the land area on Earth. It's a lot of ground to cover.

To accomplish his goal, Knies enlisted a consortium of organizations and governments through a high-powered group he belonged to—the Club of Rome. The Club was formed in 1968 by an elite group of industrialists, scientists, and economists to figure solutions for the future of humanity. Desertec fit the mold. The solar concept, which at first blush seemed outlandish, soon was taken seriously by serious backers. Advancements in solar technology and transmission helped those backers to get behind the Desertec vision.

Ninety percent of the world's population lives less than two thousand miles away from a major desert area. Electrical transmission lines can stretch that distance, putting clean, renewable energy within the grasp of most of the world. In addition to solar energy, deserts can also produce lots of wind and geothermal power. Geothermal power comes from the internal heat the planet itself produces. Hot springs are good examples of heat from the Earth's core rising to the surface. Wind power, of course, is dependent on regular breezes. Wind is created when colder air and warmer air collide. The difference in temperature creates pressure that forces air into motion.

Cold air and hot air battle it out frequently in deserts. Moisture traps heat. And dry air allows that heat to escape more quickly. Which is why at night without the sun to keep temperatures warm, it gets so cold in deserts. Daylight rapidly heats things back up. Therefore, wind arises.

Desert winds go by different names. Some describe degree. A haboob is a dust storm. Some describe direction. A shamal is northwesterly. Desert winds conjure something mystical, something psychological. In California, when the Santa Ana winds, or "Devil Winds" as they're known, blow from the desert, a wanton fix of black magic is said to take hold of people. There is a sort of despair in the air. Crime rates actually do increase, and psychological studies show that mood shifts do occur.

Maybe the "devil" association is devised to satisfy deep, sad thoughts—

thoughts of emptiness. A desert, bereft of life, isn't something we look at as if it's filled with promise. A desert is a lonely place, vast and unforgiving. Recast as a state of mind or an experience, a desert is not a place you'd ever want to be in. It's a wasteland. A place where things decay, crumble, and die. Or have we, as Knies posited, just been looking at them wrong?

Rather than think of deserts as unusable wastelands, Knies and his group of scientists, politicians, and economists at Desertec came up with a different thought: to recast deserts as the engines of the world. And what better place to start than the Sahara Desert? Only the Arctic and Antarctic deserts are bigger in size than the Sahara. In terms of hot deserts, the Sahara is by far the world's largest. Indeed, the next biggest desert, the Arabian, is less than one-third the size.

The road leading to the Noor solar plant outside Ouarzazate, Morocco—the gateway to the Sahara, as it's known—seems out of place. The road is newly paved asphalt and manicured on either side. It tracks in contrast to the desert scrub and sand gravel that continue in all directions for as far as the eye can see. A gate manned by armed guards is at the plant's entrance. Unless you are supposed to be there and have identification, you aren't getting by. It's no wonder. The Noor plant is a multibillion-dollar structure in the middle of nowhere. When it's complete, it will be the largest concentrated solar power plant in the world. It will generate so much energy that it will be able to power Europe. It is designed as a model, where world leaders, educators, professionals, and students can visit to tour the facility and learn about the power of the sun's energy.

Knies had zeroed in on the Sahara because of its insolation and location. Not only could it power local communities, but its potential could stretch across continents. Still, Knies realized early on in his Desertec gambit that he'd need to connect to an augmented power grid for this alternative energy to become more widely accessible.

A smart grid powered by renewable energy was the dream for Knies and is the goal for a lot of so-called clean-tech industry entrepreneurs. A smart grid incorporates technology into traditional transmission lines, transformers, substations, and end users. It better controls capacity and loads and responds more quickly to spikes and lulls in demand. It also distributes loads more evenly so blackouts and brownouts are less frequent.

Because the grid can integrate power generated by customers, it can take in energy from more sources and is especially accommodating to renewable energy systems. Let's say you install solar panels on your roof, but you don't need or use all the energy they produce. The excess can be fed into a smart-grid system. With increasing breakthroughs in science and engineering, we can all pretty much wrap our brains around smarter energy systems. However, despite the possibilities of connecting most of the world to renewable energy sources, expansive energy grids are scant.

Engineering thousand-mile-long power lines isn't the issue. Long power lines are problematic politically: Who gets charged or paid for what, and how much? Territorial battles ensue.

For solar power to make its way from the Sahara Desert in Morocco to Europe, it would have to feed into a grid system that would then transmit the power across the Strait of Gibraltar to Spain, where it would then connect with a smart European superenergy grid.

The European supergrid proposes to link all fifty European countries with shared power resources to create efficiencies. The efficiencies would ostensibly lower energy prices for everyone on the continent. Countries with hydropower, for example, could feed in their share, just as countries without hydropower could feed in their share from another energy source. The potluck gamut is still in the works and has been stalled by everything from infighting to vacillating oil prices and the Syrian war. But the desire to create a single electricity market for the Middle East/North Africa (MENA) region and Europe remains. Here's why: On a per-kilowatt-hour price basis, solar energy from the Sahara can be produced at less than ten cents, which is half the price of what people pay on average in Europe.

The Noor Ouarzazate Solar Plant

The Strait of Gibraltar, which separates Europe and Africa, is just nine miles wide at its narrowest point. For decades, people have looked at that space between Morocco and Spain as an opportunity to connect the two continents. In the 1920s there was a reportedly popular proposal floating around to dam the strait and build a hydroelectric power plant that would service the area. (It should be noted that the proposal was part of a bigger concept called Atlantropa that aimed to partially drain the Mediterranean Sea so more landmass would be exposed. The land, as the idea went, would become farmland and expand Europe's colonial reach. Although peaceful, the principle behind Atlantropa echoed the Nazi party's master plan of territorial dominance. Needless to say, Atlantropa never came about.)

Other ideas for the strait have included a floating bridge, a roadway, and a rail tunnel—all with the hope of artificially connecting the lands. Farther east, undersea cables are being laid. They are meant to connect Tunisia with Malta and, in turn, Europe. Malta is already connected to the European grid by deep-sea power cables.

In September 2017, Nur Energie, which operates a solar plant in the Tunisian part of the Sahara, announced plans to export electricity to

Europe using those cables. By the end of that year, it looked as though Knies's vision, which had begun thirty years before, was finally coming about.

On December 11, 2017, after a long illness, Knies passed away in his home of Hamburg, Germany. He didn't get to see his vision realized. But it's clear that at least part of his grand plan to reshape the world is on course.

Friedrich Führ, a founding director of Desertec and a friend of Knies's, put it this way: "The helpful little red rectangle on the map, which symbolized the surface area required for the power plants, became Knies's trademark. Looking at the huge expanse of the Sahara and the little red rectangle, we immediately see what numerous extensive studies and publications have now proven beyond doubt: Desertec works.

"A great visionary and realist with a sharp mind, Knies promoted a wider global debate on renewable energy, pursuing his aim of speeding up the transition to clean energy. In his mind there was no doubt that this step would have to be taken, but he feared that it might be taken too late to limit global warming to a maximum of two degrees [Celsius]. He was convinced that we have to do everything in our power to speed up this unavoidable transition. He was always open to suggestions and ready to learn something new every day. A question he frequently asked was: 'Is humanity insane enough to commit collective suicide?'"

The Noor plant and the broader sustainable energy operations taking hold around the world prove that we are not.

Abderazzak Amrani, a serious-looking and serious-minded solar engineer at the Noor Ouarzazate solar plant, dons a white hard hat and orange reflective vest before escorting a visitor onto the roof of the facility. His dark hair and olive skin are a reminder that Noor is a job provider to the local community as well as an energy hub for the MENA region and beyond.

While the plant is meant as a beacon for the world, there is also something proudly Moroccan about it. The inside flap of the booklet that details the plant's facts and figures has a photo of King Mohammed VI. And the acronym under which the plant goes, MASEN, stands for Moroccan Agency for Sustainable Energy.

On the roof, Amrani points to the different panels that stack out in rows far into the distance. He clicks off statistics: The Noor Ouarzazate plant is built in four stages, producing more than five hundred megawatts of power. When it is linked with the other Noor plants in Morocco, solar capacity will exceed two thousand megawatts of power. MASEN's goal is to have more than half the country's energy come from renewables by 2030 and be on a path toward exporting it.

Producing solar electricity is a relatively simple engineering process. Photovoltaic (PV) cells made from silicon, mirrors, or lenses capture sunlight. In the case of PV panels, semiconductors transfer electrons through circuits that can be tapped for power. Mirrors and lenses are used as concentrated solar power. They are aimed at a source of water to heat it and produce steam. The steam turns turbines that create electricity.

At Noor Ouarzazate, molten salt is mostly how the solar energy is stored. The sun's energy heats the salt, which can then be used to create steam that turns the turbines to create electricity. Salt can stay hot for up to ten hours, and provides seven hours of storage at the Noor plant.

Concentrated solar has been difficult to utilize in the past because it is direct voltage. We often hear about AC or DC in the context of electricity. Alternating current, AC, switches directions and allows higher voltage to service common appliances. Plug your toaster into a direct current, DC, system back in the day, and you'd end up looking like Wile E. Coyote after opening an Acme package from the Road Runner: smoke festooning from your fried and frizzled head.

Advances in direct current technology, where power is transmitted in one direction at high voltage, now allow loads to be stepped down for common use. For solar power to reach Europe from Ouarzazate, it would have to travel more than five hundred miles. Concentrated solar keeps more power intact over distance.

Amrani says that for the moment the MASEN initiative is to service the local grid. The power lines to it stand out. They appear in the desert like Transformer figures, hundreds of feet tall.

Mirrors, steel, and concrete are what you see in various combinations, whether they are the rows of solar panels or the engine structures. The modern campus, seven miles square in size, is a world unto itself here.

In the distance, though, is a red-hued casbah. Amrani explains that it is the future visitor center, where people can come and stay and learn about what goes on here. "This is the goal of MASEN," he says. "Not just to produce electricity, but to build a sustainable project that others can see; maybe develop their own."

But what if every country did? What if solar panels were indeed placed onto the surface areas of deserts en masse?

A team of scientists figured out what would happen: Earth would heat up and dry out and wind patterns would change dramatically. They found that the heat absorbed by solar panels cooled the deserts where they were installed and made the deserts even drier. And the energy that was captured and transferred heated up the areas where it was used. The net effect was weird climatic consequences that included pass-along changes in weather patterns.

"There are consequences involved with these processes that modulate the global atmospheric circulation," they wrote in an article for the scientific journal *Nature Climate Change*. The global atmospheric circulation is what carries heat from the tropics to the polar areas; it is what spawns seasonal weather. Desert solar absorption messes with these age-old patterns, possibly triggering droughts, wildfires, and heat waves. Expected seasonal temperature change could also be thrown off.

Paul van Son, chief executive of the German company Dii Desert Energy and an active member of the Desertec movement, says members have long since backed off Knies's original bold plan for every inch of desert to be covered with solar panels. He says a more holistic sustainable energy program is being embraced, where wind, solar, and other renewables are part of a portfolio that can power the world sustainably.

"We started with a top-down approach, but realized that would never work," he says, as he zips along the German autobahn.

The new, bottom-up approach holds out the prospects for energy synergies between geographic regions; sunlight during the day in one place is used as the major energy supplier while hydro in another region steps up supply later at night, for instance. "In the future, we'll have all sorts of suppliers," van Son says. He predicts that regions will add to the renewable energy market slowly. "But once they see lower prices, then everyone will ask for connection," he says.

Still, capturing the sun's power will be the biggest part of that offering, and deserts are already being refigured as energy spots. India, China, and the United States are investing big to exploit desert sun. They each boast about having the biggest solar farm by one measure or another, year after year. But as of this writing, none tops the concentrated power being harnessed from the Sahara.

Eventually, no matter the consequence, at least a portion of deserts in most countries will likely be covered in solar apparatus. The energy potential is just too great to ignore.

No-Blade Wind "Turbines"

 A Spanish company has devised a way to capture wind energy from turbines without blades—in other words, turbines that aren't turbines by definition.

Vortex Bladeless has patented what it says is a more cost-effective, environmentally friendly, and simpler way to produce energy from wind.

The wind generators look like giant baseball bats stood on their handles. They are forty-one-foot cylinders that capture wind through a process called vortex shredding. The cylinder oscillates when the wind blows and generates electricity via an alternator. The technology works off the concept of vorticity, or the tendency of particles, such as air, to spin at a particular point. Interestingly, it is not so much the wind itself per se that powers the "turbines," but the vibrations created by the wind's energy.

"When the wind vortices match the natural frequency of the device's structure it begins resonating, hence oscillating, so the bladeless wind turbine can harness energy from that movement as a regular generator," the company explains.

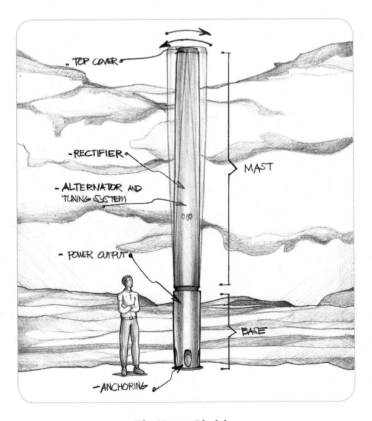

The Vortex Bladeless

A traditional wind turbine works when wind spins the turbine's blades and the turbine captures kinetic energy that is used to power its generator.

The Vortex Bladeless technology is based on fluid dynamics. Engineers and architects typically try to avoid vortex-induced vibrations because, as might be imagined, shaking buildings, bridges, and structures freak people out. Vortex Bladeless has embraced the vibrations, using them as energy sources.

Still, plentiful wind is necessary for the devices to operate. Deserts produce vast amounts of wind, which is why many wind farms are sited on desert floors.

Wind is produced when temperatures—atmospheric pressures—clash. A desert's light colors reflect the sun's energy more than your average land area and create a highly unequal environment of heating and cooling. High winds are the by-product. The same holds true for polar regions. For opposite reasons, it's also why oceans produce a lot of wind. Oceans absorb heat rather than reflect it, yet they also produce hot-air/cold-air tension that produces winds.

A big criticism of wind farms, whether desert or ocean, is their unsightliness and how much space they encompass. A single turbine, depending on its size, might need fifty acres of unobstructed space around it. Vortex Bladeless devices take up half as much space as run-of-the mill wind turbines and can come in lower-watt versions that stand only nine feet tall. The average industrial blade wind turbine stands approximately three hundred feet high.

Vortex Bladeless says the devices can be lined closer to one another than blade turbines because blade rotors need more "swept area" to be efficient. Swept area is the diameter of wind space needed to generate power.

Another issue with large wind turbines is their effect on nature. Wind farms have historically interfered with desert and marine habitats and the flight patterns of birds. Vortex Bladeless claims that without blades and given the device's smaller height, "it won't disturb wildlife and allows birds to have a higher visibility while flying." The company is working with a bird conservation organization to ensure their power-generating devices and birds can share the same wind.

Vortex Bladeless may have not only reengineered how wind energy is captured but also how deserts may look in the future: Commercial pinwheels, be gone.

Ocean Energy

 Deserts may be great sources for solar and wind power, but oceans hold big energy potential as well—perhaps even more than any renewable energy on land.

One recent study says a massive wind farm in the North Atlantic could power the needs for all of humankind. There is more open space in the oceans, and winds can pick up more speed than on land.

Offshore wind farms are nothing new. It's building them and connecting power lines that are troublesome, given the rough and salty conditions. Dive deeper, though, and new forms of ocean power can be had: ocean thermal energy and wave energy.

To produce energy, ocean thermal energy utilizes the differential between the cold water on the bottom of the sea and the warm water at the surface. When warm surface water is siphoned from the ocean, it heats a fluid, such as ammonia. The fluid's vapor blows through a turbine attached to a generator. The generator produces electricity. Once the fluid has done its job of vaporizing, it flows to another apparatus, where it is cooled by the deep water that has been siphoned from ocean depths. The ammonia or other fluid surrogate never leaves the system. It just acts as a conduit.

Some floating ocean-thermal-energy platforms look like a man-made octopus; pipes float down from the body of the platform where the heat exchanger and other mechanics are housed.

Several companies have begun operating ocean thermal plants. The world's largest plant, developed by Makai Ocean Engineering, began operating in Hawaii in 2015.

Japan has a demonstration project in Okinawa that is being held out as a model facility. One hundred kilowatts of power are produced by the plant, and public tours can be had to preview the future of what ocean energy might look like.

Other plants are being developed in other locations around the world, from Asia to the US East Coast.

Wave energy is also being harnessed. There are several different types of wave-energy devices:

- Terminator systems use a chamber or pipe that extends perpendicular to waves. When the water enters the opening, it is trapped. Its up-and-down motion acts like a piston to create energy.
- Attenuators ride along waves and use pumps to capture the energy of the wave as it breaks across.
- Point absorbers ride waves like a buoy and utilize the bobbing to generate power.
- Overtopping devices essentially are floating dams just above the surface that use the pressure buildup of ocean water, forced by waves breaking over the top, to generate power.

Globally, wave power is on the rise. Wave farms, sometimes called wave parks, are operating off the coasts of Portugal (the world's first), Australia, Great Britain, and the United States.

Soon, oceans may be teeming with more than marine life. Power plants are headed out to sea.

CHAPTER 5

Cool Roofs and Roads

Las Vegas is where humans announce their presence most to the universe. The Strip bills itself as the brightest spot on Earth. Astronauts on the International Space Station can see it prominently from outer space. They've even taken photos to prove its ultimate brightness against the dark surrounding desert terrain at night.

The brightness, of course, doesn't come from the sun. More than a billion watts of electrical power make the concentration of light from Sin City's hotels and casinos illuminate so magnificently. It's where the famous Sky Beam, the world's strongest light at 42.3 billion candela, blasts out from atop the Luxor hotel and shines vertically with the help of thirty-nine xenon lamps. It's where the sparkling facade of the Mandalay Bay casino acts as a forty-three-floor wall of luminescence, and where the Caesars Paris Las Vegas resort boasts a glowing half-sized rendition of the Eiffel Tower lit from bottom to top. All of this is attributable to human building and engineering. To some, though, it is perhaps the ultimate desecration of nature's elements.

Two hundred and fifty miles south of Las Vegas, in Yuma, Arizona, the sun produces its own record feat of accomplishment. There, the sun shines more than 90 percent of the time, amounting to more than four thousand hours of direct sunlight per year. Most cities get around half that much. Yuma, then, by the sun's measure, is the planet's brightest place.

Yuma's attraction to the sun is its lack of clouds. It has on average 242 days of clear skies per year. Clouds help bounce the sun's rays back into outer space, much like polar glaciers or other white areas. The dark ground or the blue ocean do the opposite, and attract and absorb the sun's energy. This is known as the albedo effect, or the amount of solar radiation reflected by the Earth's surface. Ice has a high albedo, whereas soil, because of its darkness, has a low albedo and stores lots of energy. Yuma's sunny and cloudless days and inland location make it the perfect target for sunshine to soak into terra firma.

Yuma is a 121-square-mile flat patch of earth that displays itself to the sky above in a multitude of brown hues: Amber, beige, chestnut, and chocolate appear most to the naked eye. It's those tertiary colors that are just dark enough—and on their way to black on the color scale—which can keep in vast amounts of heat.

Cities around the world are increasingly mirroring Yuma's heat effect. Clear skies and topsoil aren't the causes for their heating up, however. Man-made blacktops are the reason.

Dark areas are growing and outnumber white areas inside most city boundaries.

The sun's energy is captured in blacktops, and at best just 20 percent of that radiation escapes, making its way back up into the sky and eventually into outer space. The rest of the energy stays trapped in the ground or just above the surface, heating the material in which it is stored and jacking up temperatures in the vicinity. A lighter-colored surface, on the other hand, would bounce more than three times that amount of heat back up into the atmosphere, keeping ground temperatures far cooler.

Around 60 percent of city surface areas, on average, are composed of black or dark colors creating artificial hot zones. While only 3 percent of all the land surface in the world is made up of cities, urban areas are expected to triple in size by 2030. That means a lot more artificial heat: islands of it.

Urban heat islands, as might be imagined, are a huge problem. The concentrated heat not only promulgates global temperature rise, it exacts pollution and damages public health. When tailpipe emissions are heated up, for instance, they become smog. Breathing smog causes respiratory problems and asthma attacks, among other maladies. By example, people living within five hundred feet of freeways in California, where traffic is legendary, suffer higher rates of asthma, heart attacks, strokes, lung cancer, and preterm births.

Combine the proliferating reach of urban sprawl with the fact that more of the global population now lives in cities than outside them for the first time in history, and the health and wellness recipe for people and the planet isn't good. "They paved paradise and put up a parking lot," from Joni Mitchell's 1970 song "Big Yellow Taxi," is an apt verse of what's happening in the world today.

Urban heat islands can make a city center as much as twenty degrees hotter than its surrounding suburbs. New York City could end up looking like a dystopian nightmare because of the urban heat effect combined with global temperature rise. *New York* magazine has described what havoc heat rise might wrought: famine, economic collapse, heat death, plagues, unbreathable air, perpetual war—in short, doomsday.

David Wallace-Wells, in his *New York* article "The Uninhabitable Earth, Annotated Edition" says, "It is, I promise, worse than you think. If your anxiety about global warming is dominated by fears of sea-level rise, you are barely scratching the surface of what terrors are possible, even within the lifetime of a teenager today. And yet the swelling seas—and the cities they will drown—have so dominated the picture of global warming, and so overwhelmed our capacity for climate panic, that they have occluded our perception of other threats, many much closer at hand. Rising oceans are bad, in fact very bad; but fleeing the coastline will not be enough.

"Indeed, absent a significant adjustment to how billions of humans conduct their lives, parts of the Earth will likely become close to uninhabitable, and other parts horrifically inhospitable, as soon as the end of this century." His article is aimed, of course, largely at residents of New York. A country away, however, in Los Angeles, such a bleak future is being painted over. Literally.

"We're effectively re-covering the surface areas of streets and roofs," says Jonathan Parfrey, who helped pioneer a citywide initiative known as the "cool roof" law. The idea is to re-coat black surfaces—playgrounds, parking lots, alleyways, and rooftops; streets, sidewalks, and basketball courts—and make them more reflective, thereby cooling local temperatures.

Parfrey is the executive director of the L.A. nonprofit organization Climate Resolve, which looks to foster local solutions for climate change in the hopes they will catch on in other cities across the globe. Cool roofs, he hopes, will be embraced worldwide.

Parfrey is in his fifties. He grew up in L.A., and he witnessed the downtown area revitalize and thrive as a social hub. As a commissioner for the Los Angeles Department of Water and Power, he became all too familiar with the effects of climate change: strain on the power grid, public health and welfare concerns—and urban heat. When he learned about the promise of cool roofs, he pushed for it. It seemed to make a lot of sense and simply could make the city a better place to live.

"I saw it as a win-win. The city gets cooler and we'd use less fossil fuel for air conditioning," Parfrey says.

In 2013 Los Angeles became the first major city in the United States to mandate that all new homes use roof materials that reflect sunlight. There are plans for wide use of reflective roads, as well. The goal is part of a strategy to reduce the heat temperature of the city by 3°F by 2035. That's a lot, considering that global temperature rise is expected to jump 3.6°F by that time.

The cool roofs process applies to new and existing roof surfaces on houses, garages, and commercial buildings. Low-sloped roofs and steep-sloped roofs reflect the sun differently and have different requirements.

Most houses have steep-sloped roofs, whereas commercial buildings typically have large, flatter-surfaced roofs.

In general, a larger, flatter surface is more predisposed to attract and absorb more heat, but reflectivity all depends on the geography and the angle of the sun. Builders and roof workers have to take all of these considerations into account to best cool a roof. And they do. And they are cooling them.

The city is on track to install thousands of cool roofs throughout the metropolitan area, and L.A.'s mayor, Eric Garcetti, hopes this will set the green standard for urban areas of the future.

In addition to reducing local temperatures by more than five degrees, when combined with tree shading, cool roofs can also save water. Researchers at the Department of Energy's Berkeley Lab discovered that if cool roofs were adopted throughout Los Angeles County, eighty-three million gallons of water that would have gone for irrigation would be saved. That's just L.A. And that's just water.

There are seven cities around the world the size of Los Angeles or greater. If just those cities alone cooled their roofs, it could conceivably save more than a billion gallons of water, which is a phenomenal sum— enough, for instance, to provide water to every resident of Los Angeles for several days.

Fresh water is a big concern in Los Angeles, as the city struggles on and off with long-term droughts. Beyond water, if all urban areas joined L.A. and adopted cool-roof policies, it would remove forty-four gigatons of greenhouse gases from the atmosphere, which is the amount that three hundred million cars emit over twenty years. All that gas produces excessive heat. And in cities like Los Angeles, the number of extremely hot days is expected to triple by mid-century.

Heat rise comes at a big cost. By 2100, urban heat is projected to cost cities 11 percent of their economic output, due to medical care, labor, and other losses.

A future city built with cool roofs and reflective roads could thrive, ensuring health and prosperity.

The Global Cool Cities Alliance is a group that is looking to get more cities on board with the cooling approach. To that end, it has devised a rendition of what a cool future city might look like versus a city with dark roofs and surfaces. The difference is stark.

In the cool city, gardens grow atop buildings whose roofs are painted white. Bicycles glide by on gray streets. The air is cleaner and cooler. Even on an eighty-degree day, people can enjoy dining outside on sidewalks.

The city with dark roofs, meanwhile, experiences heat warnings, cars emitting pollution, and people sweltering inside buildings. Air conditioning is blasting in the buildings that have it. Coal plants are on overdrive to produce electricity. The message is clear.

The C40 is yet another group of people aiming to make cities more livable in the future. Its initiative connects ninety-four of the world's largest cities, representing seven hundred million people, with the goal of sharing best practices. It, too, has embraced cool-roof principles. The effect that could be made by all these groups urging city officials to brighten their dark spots could radically alter the face of the Earth; it would be like sunblock for the planet.

A cool roof or reflective road isn't just something that is lighter in color than asphalt, although color is important. A single black surface in a downtown area can reach 150°F or more on a summer's day. A clean white roof will keep temperatures on it 50 degrees cooler than, say, a darker one. And that's simply just the power of color.

Special reflective pigments can be mixed together to protect the roof from ultraviolet light, and cool it, as well. The tiny reflective particles not only block radiation but also make surfaces tougher and more resilient. Inside, building temperatures are reduced enough so that a building might not even need air conditioning.

The most advanced form of cool roof utilizes photovoltaic panels.

These are otherwise known as solar panels. Energy absorption comes into play here, spitting out renewable energy.

By concentrating on cooling roofs, as much as 15 percent of a building's energy use can be saved, resulting in billions of dollars in savings a year if all US cities lightened their loads.

There are other surfaces that can help cool things inside city boundaries, too. Chicago has launched a Green Alley Program to mitigate its urban heat effect. Singapore is laying out cooler pavement. And other cities are examining their blacktops to lighten basketball courts and parking lots. Los Angeles, though, is hoping to create what might be the world's first entirely "cool" neighborhood by focusing on roads.

There are fifteen pilot programs for reflective roads throughout the city of Los Angeles alone—one for every city council district. "No other major city is taking this on. That is what's so exciting," Parfrey says.

On an overcast winter's day in Canoga Park, a neighborhood in the West Valley of Los Angeles, Parfrey takes a gun out of his jacket pocket and points it at the ground. It's a thermometer gun that gauges precise temperature readings. He finds that the sidewalk is 65°F. A few steps away, he points the gun at black asphalt. There the ground is 68°F. Just a few steps from that spot, the asphalt is covered with reflective material. The temperature there is 63°F. The difference is remarkable.

"The neighbors like being part of this test process," he says. Lining the street are two-story apartment buildings. Kids ride their bicycles up and down the sidewalks. Several SUVs packed with families and pets pull away from the curb.

Reflective roads can remove the scorch of tar on bare feet or dogs' paws. Kids can be more active in summertime. In general, a cooler environment makes it more pleasant to be out and about. This is important in high-temperature zones like the West Valley. The Valley gets (like) super hot in summer. For days it can remain above 100°F, and the temperature difference between it and beachside of the Santa Monica Mountains can be nearly 20 degrees. At night, reflective roads help, too. "You use less electricity," Parfrey notes, because streetlamps don't have to use as much energy.

A cylindrical tank on a truck disperses the light gray paint, and crews of workers spread it across the street, from curb to curb, until only a thin coat remains. That's all it takes. Shadows of trees and telephone poles and power lines form where the dark lanes once were. Patches of clear sky that let the sun shine bright sketch these silhouettes more distinctly on the reflective surface. Unfortunately, tread marks dull the material's effect, which is why Parfrey is lobbying hard for Canoga Park to become the first whole neighborhood in the world coated in reflective paint. That way the smudges from asphalt on the older dark roads won't be so distinct on the new reflective roads; the smudges will be lessened with more light roads to drive on. As it stands, the test road is only a few hundred feet long, bordered by black tarred intersections.

Parfrey's little temperature gun experiment occurred in winter on an overcast day. Had comparisons been done in summer, the differential likely would have been even greater between the dark surface and the lighter reflective coating. The coating comes with an additional benefit: It makes streets more resilient. The main contributor to roadway deterioration isn't traffic, it's sunlight, Parfrey says.

Not all of the sun's light reaches the ground, even on cloudless days. One percent of the sun's radiation is immediately trapped by the upper atmosphere before it can make its way closer to the Earth's surface. Another 20 to 25 percent of sunlight (technically solar irradiance) gets lapped up in the troposphere by greenhouse gases. Some 30 percent of the sunlight that finds its way farther down gets bounced back by high albedo surfaces—polar ice caps, desert sands, and such. That leaves about 50 percent of the solar energy remaining to finds its way into land or oceans. But even when that energy gets absorbed, it doesn't stay forever. Eventually this energy also releases back into the atmosphere and into outer space. It becomes part of the Earth's own thermal emission.

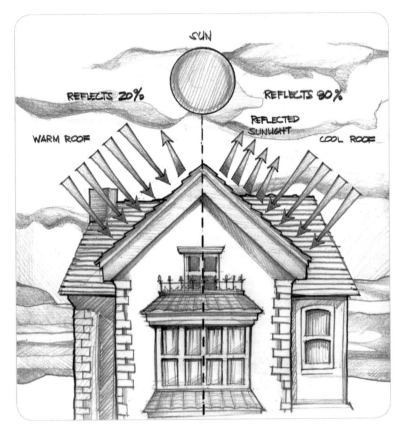

A Cool Roof

The difference between how much solar energy gets trapped on Earth and in the atmosphere and how much leaves is what provides a livable temperature for the planet. Currently our global temperature rests at about 59°F. Which is why when more of the sun's energy is absorbed in dark materials, global temperature rise is produced. Reflective materials can help sustain temperature balance.

While cool roofs and reflective roads seem like rather innocuous feats of engineering (after all, lighter surfaces dot communities in warmer environs such as the Mediterranean and the Caribbean for a reason; those communities long ago figured out the cooling power of light colors), their climate effects may not be as brilliant as they appear.

Stanford professor Mark Jacobson released a study of what happens when roofs are painted white. He found that instead of cooling the atmosphere, reflective surfaces actually warm global temperatures. Jacobson's data show that local temperatures are indeed cooled by installing more reflective coatings and the urban heat island effect is dampened, but the Earth overall is warmed.

Essentially Jacobson found that the solar radiation that was dispersed from the ground on a local level became absorbed by aerosols higher in the atmosphere. These aerosols add to the mix of greenhouse gases and warm temperatures on a global scale. Moreover, his analysis indicates that increased reflection exacerbates pollution. "More people will die," he declares, if cool-roof projects are adopted on a grander scale in certain regions.

Instead of cool roofs, why not just install solar roofs? Jacobson wonders. At least the energy absorbed is made into something productive—electrical power—he says.

To be sure, Jacobson's study has received a fair amount of criticism within the scientific and environmental communities. One of the biggest critiques was the exclusion of energy savings that cool roofs afford and, in turn, their carbon emissions. Lower carbon emissions could offset the extra global temperature rise Jacobson discovered running urban heat island and cool-roof models. Jacobson actually raised this point directly in his study. "The local cooling due to white roofs may reduce or increase energy demand and thus other emissions as well, a factor not accounted for in these simulations. This feedback should be considered in any final assessment of the effects of white roofs on climate," he wrote.

And while it's true that cool roofs reduce building temperatures in summer, and therefore can reduce energy demands, building temperatures in wintertime are also reduced, and therefore energy demands for heating could rise. There is four times more demand for heating than cooling, according to Jacobson. Hence his call for more analysis taking into account energy heating and cooling factors when it comes to cool roofs. He repeats during the course of a long conversation his advocacy for solar panels that produce a cooling effect and at the same time energy.

By the sun's accounting alone, cool roofs increase the albedo effect and lower local temperatures. But man-made materials and energy interfere with that in ways that are unknown, like messing with cloud formations, or more general global warming.

"We need more tests," Parfrey agrees. He says the mainstream of the scientific community believes cooling works, but there may be an even better solution out there. "I'm agnostic. If there is a better solution, I am all for that," he says.

Meanwhile, until something better comes along, Parfrey is pushing ahead in hopes that Canoga Park will indeed become the world's first artificially cooled neighborhood.

After just a few months of use, the reflective road in Canoga Park is littered with skid marks. Like most any other street, there are trash, leaves, and a trail of cigarette butts on it. The fact that the street is special is almost lost until the sun shines. Then, the luminosity of the material appears and the road brightens. It's cool.

Smog-Fighting Shingles

 Although its name was coined only a little more than a hundred years ago, smog has been around for centuries. It is a mix of smoke and fog with some help from the sun's energy.

When warm moist air gets trapped low to the ground and circulates over cool surfaces in a process called advection, fog is born. Add smoke to the mix, and you have smog.

The smoke from forest fires or coal burning creates smog. Vehicle exhaust fumes produce a different kind, called photochemical smog.

This happens when sunlight mixes with nitrogen oxide and other pollutants from exhaust pipes and forms a brown haze. In cities and other areas of concentrated pollution, this is especially problematic because it endangers public health. Smog, when inhaled, can produce a deadly assortment of diseases and respiratory ailments. Asthma is a common result, and lung cancer can also be caused.

While the pollutant is primarily formed from emissions from combustion engines—cars, trucks, and buses—power plants and heavy machinery can cause smog, too. Smog is a leading cause of acid rain, which destroys the ecosystems of lakes and forests.

As more people migrate to urban areas, smog promises to get only worse. Already, a shocking 91 percent of the world's population lives in places where air quality exceeds World Health Organization standards.

There isn't much room for more people to be put in danger. Which is why an invention by one of the world's biggest conglomerates is so important: It can repurpose potentially deadly smog into cleaner, fresh air.

The company 3M, best known for its tapes and adhesive products, has figured out a way to suck smog out of the air before it has the chance to do more harm. By mixing what 3M calls "smog-reducing granules" into common roof shingles, hazardous pollutants can be transformed into ions that improve air quality. It works when the sun's rays come into contact with the specialized shingles. This sets off a chemical reaction process that can turn any roof into a smog-reducing machine. The roofing granules are designed with a specialized photocatalytic coating. "As sunlight hits the shingles containing the smog-reducing granules, radicals are generated and transform nitrogen oxide gases into water-soluble ions improving air quality," 3M explains.

Los Angeles and Beijing are famous for the smog that haunts them. In the past, the cities have attempted different schemes to reduce traffic congestion and lift air quality. But smog persists. Smog-reducing roofs could go a long way toward clearing their skies. While one roof might not have much of an effect on citywide smog, a community of them would.

Smog is closely linked with the urban heat island effect. Blacktop surfaces absorb more of the sun's energy, increasing local temperatures and heightening the chances for smog to appear. 3M may have figured out a way to harness the sun to produce fresh air.

Cool Alleys

Chicago has more alleyways than any other city in the world. Those dark places attract and absorb sunlight, spiking urban temperatures several degrees higher than in the suburbs. On some hot summer days, Chi-town's downtown center can be as much as ten degrees warmer than its suburbs, making it a famous urban heat island.

To repel the sun's energy, the city developed and has launched a Green Alley Program. Chicago is resurfacing its nineteen hundred miles of alleyways with lighter, sun-reflecting materials that are also permeable. (Chicago's alleys aren't connected to its sewer and stormwater systems, causing chronic instances of localized flooding.) At the same time, the Chicago city government is encouraging residents to plant gardens, trees, and install green roofs, all in the hopes of cooling hot urban temperatures.

Higher temperatures cost a city economically not only because of higher energy costs (for cooling). The heat also impairs air and water quality and compromises public health.

One study predicts that global temperatures will cost cities twice as much as their suburbs and more rural communities because of public health costs and labor losses.

Chicago employs several different techniques in order to green or cool alleys, depending on the location, width, and grade of the alley itself. Different types of materials, including recycled concrete, rubber, and permeable pavement, can be used. Effectively, the process involves either entirely replacing alleyway surfaces or recoating them. The construction details can get involved, but the goal is to redesign how the city looks from above.

As it is, Chicago, birthplace of the skyscraper, is a dense city of 2.7 million people living on 237 square miles that is a grid system. Grid cities keep heat trapped, especially at night, because they are built in clusters, whereas cities that wind around, such as Boston, allow heat to more readily escape. Chicago's location on (dark-colored) Lake Michigan and an expansive cityscape that rolls out flat into the fields of the Midwest make it all the more catchy to the sun's attention. At night, it even produces its own sort of glow—sky glow.

A dark-sky project will see "dark sky" lights replace traditional street lighting to direct light downward. Darker night skies will bring the city back a little closer to a time when we could see the stars at night without the interference of blue light, the type most artificial lighting produces. Blue light also interrupts the nighttime habits of wildlife and may cause negative health effects on humans (disrupting our circadian rhythms).

Undoing urban development to get back to how nature once was isn't an option, of course. But lightening the effect of construction and building can make cities such as Chicago less appealing to the sun.

HUMANS V.
LAND & OCEANS

CHAPTER 6

Smart Soils

Nature's nuclear bomb is underfoot. Here in the Indonesian rainforest, the world's largest reserve of tropical peat carbon is stored. It contains seventy billion tons of carbon locked up in a mash of dark brown material—the remains of plants partially decayed for millennia and sopped in water—swamps, marshes, and jungles. Peat contains more carbon than any other soil type. If all the carbon was released from this peat moss, it would cause an explosion of global temperatures and pollution. It would be like the entire world burning two years' worth of fossil fuels all at once.

Indonesia's peatlands contain as much as nine thousand tons of carbon per hectare. One hectare is equal to 2.5 acres. And there are 13 million hectares of peatland in this country, spread out thick in its tropical rainforest. Other countries outside the tropics have even larger areas of peatland than Indonesia. But what makes this nation unique is that it is degrading its peatlands at a massive rate, arguably causing the single most devastating attack on soil in the world. When soil falters, so does civilization.

Soil has many essential functions: supporting plant and animal life; filtering air and water pollutants; capturing and storing elements such as carbon, nitrogen, phosphorus, and other materials; and acting as a foundation for vegetation, as well as man-made structures. The soil in

Indonesia is being corrupted for nonessential reasons: So we can eat fried foods and brush our teeth.

Vast tracts of the rainforest in Indonesia are burned and the soil is tilled to plant and harvest palm that gets turned into oil. Palm oil is a main ingredient in toothpaste, French fries, potato chips, doughnuts, and fast foods the world over.

Every year throughout the island nation, hundreds of thousands of hectares are cleared, turned over, and converted into farmland. The practice makes Indonesia one of the biggest carbon emitters in the world, because when soil is degraded, it releases carbon into the atmosphere. If the practice continues for the next one hundred years, we'll lose as much carbon into the atmosphere as it took twenty-eight hundred years to sequester in the ground.

In 2015, when fires broke out in Indonesia, the fires burned for months, casting a haze over almost all of Southeast Asia and sending more greenhouse gases into the atmosphere in a single thirty-day period than the United States does. That amounts to sixteen million tons of carbon dioxide every day.

The average car emits less than five tons of CO_2 in an entire year. So imagine what sixteen million tons is like: a huge pileup of pollution, highways long.

International organizations such as the World Bank, and nongovernmental organizations (NGOs) such as Greenpeace, have called on the Indonesian government to set new policies to stop palm oil farming, stop the burns. In a report, *How the Palm Oil Industry Is Still Cooking the Climate*, Greenpeace detailed the ruinous practices palm oil farmers undertake. Provocative images show fields burning and people and wild animals suffering because of all the air pollution. "Companies connected to the global palm oil market continue to clear forests and are responsible for other environmental and social harm," Greenpeace wrote.

And while carbon emissions are a huge issue, the harm to soil should not be ignored. Farmers planting palm for oil typically ravage the land, yielding as much as they can from their crops and then moving on to other tracts. They don't farm for serial growth. They don't come back. In their wake is farmland left barren.

It is not a sophisticated operation, this oil palm business. On the island of Borneo, split between Indonesia, Malaysia, and Brunei, isolated workers miles into the remote rainforest ignore a visitor and continue to illegally fell trees. They drag the timber with chains and backhoes to the river and send logs downstream. The work is punishing. There are no facilities. They live in soiled tents. It's hours by foot to reach a road and hours more to the nearest community. The laborers are surrounded by jungle and ground so thick with mud that it's difficult to walk, never mind hike, never mind haul timber. The heat and humidity make the conditions all the more oppressive.

On a slope where the slog to the camp gets really tough, a lone tree stands tall at the base of the path. It beats off the sun, its leaves a healthy green, its branches strong, its trunk solid. It has avoided the ax because of its size and strength. It would be too difficult to cut down and haul away. Too much work for this ragtag group of laborers. Soon these six young men in their dirty clothes and muddy shoes will have cleared as much timber as they can to sell. Then they will torch the place. The proud warrior tree will go up in flames. He'll turn to ash, and his roots, likely there for centuries, will be ripped up. Only his dust will remain, and that will be plowed over when the soil is tilled. Oil palm will replace him.

Across the South China Sea on the island of Sumatra, a fire comes into view from a helicopter above. There are two people fanning the flames. An executive from one of the world's largest pulp and paper companies points to the fire and the surrounding area. The rest of the land, dozens of acres, has been scorched. When the farmers clear the land by burning, everyone loses—timber companies, too. There aren't proper fire controls, the executive says, so conflagrations ensue, as happened in 2015. Forests are lost, pulp for her company's products is lost, animals are displaced, people suffer health problems from the pollution, and more.

Even when there are controlled burns, the farmers don't look to the future and take care of the land properly. "We have to educate them," she says. Timber and palm oil bans don't work. People need to eat. They sacrifice long-term prosperity for more immediate basic needs. Money. Food in their stomachs.

When the two people below see the helicopter, they take off into the woods. It is difficult to make out any description of them. They are dark figures on the run, and soon they disappear from sight. But the fire still burns.

Blame for land loss and soil degradation can't be placed only on Indonesia and uneducated workers. Conventional farms do damage, too. Farms in Africa and China top the list of soil ruin because of the emphasis on short-term yields as opposed to long-term sustainability. To feed rapidly growing populations, farmland is stressed. Growing single crops—or monoculture, as is the proclivity of large, industrial farms—is also a culprit. And this occurs throughout the Western world, as well. Without crop variety, soil is deprived of a balanced life and begins to die.

Iceland at one time was covered with lush forests and had rich soils. Then the Vikings came and settled. They farmed and logged. The population grew and its needs evolved. Now, about half of Iceland is severely eroded and desertification has taken its toll on the island. It's a case in point of what occurs when human activities exploit nature without regard for the Earth.

It's the same question the author Jared Diamond asks in his book *Collapse*, when the last tree was chopped down on Easter Island: What were they thinking?

What were *they* thinking? What are *we* thinking? We are not.

At the current rate of degradation, all the world's topsoil could be gone within sixty years. And it takes one thousand years to naturally generate just three centimeters of healthy, new topsoil. That means arable land, the place where we produce most of our food, is quickly diminishing. The frightening prospect of food scarcity is becoming all too real. Farmers aren't carefully tending to the land. Overuse of artificial additives such as pesticides and synthetic fertilizers serve only to deplete soil of its nutrients over time. The Earth's fertile grounds are being made into unproductive dirt.

Thankfully there is a growing movement to bring farmland back to life. With the use of smart soil, in one growing season farmers can accomplish what would take nature a millennium.

Smart soils are compounded to create ideal growth environments for crops. This entails rotating vegetation, managing irrigation, composting, using cover crops as fertilizers, and minimizing soil disturbances.

Smart soils crack the code of traditional farming, and crowdsource the best practices farmers can use to yield the most of whichever crop they grow.

Agrointelli's Farm Mapping

Jørgen Olesen, a Danish scientist who specializes in adaption of agricultural systems to climate change, says he has always been intrigued by sustainability issues. At work on a heady project involving nitrogen flows some twenty years ago, he discovered that organic matter in soil was a problem. Nitrogen plays a critical role in agriculture because it is a primary nutrient for plant fertility. By analyzing its flows, Olesen could better estimate crop yields. But he found something else—that soil organic matter was disappearing in large quantities in the farms he was studying across Europe. More research revealed that the same epidemic was occurring at an alarming rate across the globe.

"It was a bit of a surprise," Olesen says as he sits in his office at the Aarhus University Foulum research center in northern Denmark. Bespectacled and dressed in a black T-shirt, three-quarter-length pants, and sandals, Olesen sports a small, more-salt-than-pepper mustache that matches the color of his short-cropped hair. He speaks in bursts, his Danish accent drowning out syllables. "We expect this [decline] to happen when we convert from a more diverse, grassland system where we've got a lot of that sort of soil organic matter. But that it is still happening after maybe a hundred years or so . . ." He let's the words fade. The empty air has meaning—astonishment. Soil was not coming back to life. Organic matter had gone missing.

Soil organic matter is composed of dead plants and other decaying organisms such as those found in such large quantities in peat. Organic matter is key to soil fertility and helps to effect one of the most amazing cycles in nature: From death and decay come life-producing materials that incubate seeds and help vegetation grow.

Decomposing organic matter breaks down into minerals and micronutrients that feed plants and stimulate root growth. Roots grow into plants that defy gravity and stretch, breaking through the surface of the soil where they meet sunlight. Transpiration begins. No soil organic matter and no fertile soil. No fertile soil and no agriculture. No agriculture and our current food supply quickly disappears.

The cause of soil organic matter's disappearance, it turns out, is us. When we build, deforest, till the land too much, or disrupt the landscape with vast amounts of artificial ingredients such as synthetic fertilizers, we degrade an ecosystem's biological activity that allows organic matter to do its job. The Food and Agricultural Organization of the United Nations (FAO) is very aware of the organic matter problem and is trying to take steps to educate farmers about the benefits of regenerating healthy soil.

"Any form of human intervention influences the activity of soil organisms . . . and thus the equilibrium of the system. Management practices that alter the living and nutrient conditions of soil organisms, such as repetitive tillage or burning of vegetation, result in a degradation of their micro-environments. In turn, this results in a reduction of soil biota, both in biomass and diversity. Where there are no longer organisms to decompose soil organic matter and bind soil particles, the soil structure is damaged easily by rain, wind, and sun. This can lead to rainwater runoff and soil erosion," the FAO says in a report on sustainable food production. It doesn't mince words on the severity of the potential devastation: "Severe soil erosion removes the potential energy source for soil microbes, resulting in the death of the microbial population and thus of the soil itself."

An Earth without soil is an Earth whose terrestrial life will cease to exist.

After his finding two decades ago, Olesen set an ambitious goal: to change the way in which the world farms so healthy soil can be maintained or even recovered. He developed the aptly titled SmartSOIL program, which is being adopted by the European Union and whose methods are spreading throughout the world. It relies on science, innovation, and precision agriculture, and can be helped along with a bit of high tech: remote sensors, nanotechnology, artificial intelligence, and robotics. They're the best agriculture advancements humankind can offer. After all, the stakes are high: our future existence.

Olesen's big idea is to step in and surgically fix the condition of soils. And just as in surgery, the best way to fix something is to use the patient's own parts. In the case of soil, the parts we are speaking of are nature's elements: clover and grass for cover and nitrogen; larger spacing of crop rows; less tillage; and more time between crop rotations. Natural stuff. The system, however, can be helped along by synthetic fertilizers and pesticides. To be sure, this last bit is controversial terrain. The organic farming movement finds synthetics anathema to healthy growing and eating. But Olesen says this is naive. "Too much nitrogen from plants is just as harmful," he notes. By example, synthetic leaching is lower than leaching by fava beans. "So whatever is helpful, I am for," he says. Leaching is when contaminants make their way from the soil to water sources.

Nitrogen is necessary for plants to grow, of course. But it is also a greenhouse gas in the form of nitrous oxide, which is three hundred times more harmful than carbon dioxide. Moreover, it stays in the atmosphere nearly three times as long as CO_2 and, as such, is a major contributor to ozone layer depletion.

Soil is complicated, and there are many factors in the growing process. Weeds, the amount of sunlight, temperature, precipitation, the parent material from which the soil was born, and the tools used to farm are just several of the many variables that affect yields. Just because something is labeled organic does not mean that it is better or healthier or more productive. Nature can produce its own raft of problems and diseases. The SmartSOIL common cause is to make soil as productive as it can be without sacrificing its long-term fertility.

Soil is actually a harbinger of life on Earth. It is affected by climate and living organisms and is bounded on its surface by air and at its depths by rock. The Soil Science Society of America defines soil, in part, as "The layer(s) of generally loose mineral and/or organic material that are affected by physical, chemical, and/or biological processes at or near the planetary surface, and usually hold liquids, gases, and biota and support plants." Like us, soil evolves over time and comes in different types.

There are thirty groups of soil identified by major geological organizations. The soil groups range from sandy soil types, called arenosols, to clay-rich soil types, called luvisols. Not all soil is good for growing. Take peat soil. It's too loose and thick with organic matter. To make it plant worthy, it has to be tilled, and other, more mineral-rich soil has to be added. Even then, besides the aforementioned oil palm trees, only some root vegetables can be grown from it.

It takes just the right mix of organic matter to make conventional farmland so. Surprisingly, organic matter only makes up about 3 percent of all sown land. This is where smart soil comes in.

Since we humans began mucking with the Earth to force it to produce food on demand—what we call farming—there has been a soil-management conundrum. The earliest farmers originated in the Fertile Crescent, that area of the Middle East which spans (like a crescent) from modern-day Turkey to Egypt. There, about ten thousand years ago, people gave up the nomadic life of hunting and gathering for a bit more food security—growing crops such as wheat and other grains to feed livestock. The land, of course, didn't cooperate on its own, so human ingenuity forced it. Livestock manure was used as fertilizer. A growing cycle, or ecosystem, was realized, and so began the long road to settling down. After 200,000 years of wandering, picking, plucking, and killing based on seasons, cycles, and whatnot, the first modern human beings had taken nature into their own hands. The agricultural revolution began. Put another way: Humans started geoengineering the planet by reconfiguring soil. And so it went until about a hundred years ago. By that time, farming was the major way of life. Most people lived and worked on farms. The global economy relied on agriculture. Agricultural commodities topped global trade. Then the Industrial Revolution took over.

Today, we have machines that produce food for us with seemingly no abandon. And when the soil tires, we replace it with artificial materials such as pesticides and petroleum-based fertilizers. But soil can only be propped up by us for so long. Then it gives out. This is the point we are at.

Olesen explains: "Depleting soil organic matter is mostly a consequence of having too little organic matter going back to the soil. You've been overexploiting the biomass that you've been putting in, or that you have there. You have the grain, you've removed straw maybe. This is particularly the case as you go to developing countries because that straw is used also for feeding livestock. It's maybe used also for household eating purposes and so on. They even use manures . . . nothing goes back." He says this isn't just the developing world's doing. "If we look at our industrialized systems, a bit of the problem that we've had with some of the synthetic fertilizer is when you apply these, they of course boost productivity and so on. [But] they actually tend to reduce the amount of root biomass. You also, with synthetic fertilizers, end up having relatively little below ground. If you then remove everything aboveground, you have a problem." Biomass depletion means food crops run out.

A smart soil program works to feed just the right amount of water and nutrients to the soil based on a particular growth climate. At the Foulum research center where Olesen conducts his experiments, acres of test fields are mapped, surveyed, and monitored. Each patch of land is a case study.

There is a clover experiment. Clover contributes bionitrogen to soil. But how much and under which conditions is it optimum?

There is the till versus no till experiment. Different depths of tilling are analyzed to calculate which offers the best production value.

There is the biomass experiment. The goal is to double productivity while examining the effects of annual versus perennial weeds.

And so on.

Each square area is manicured and meticulously maintained. Flags, posts, sticks, and charts indicate different happenings. It's an outdoor laboratory more than a hundred acres in size. And the landscape is something to behold.

Fields of bright yellow canola and wild flowers. Rolling greens and rich brown squares of raw soil. There are patches of forest with mature trees standing sixty, eighty, feet tall. Farmhouses and barns are in the distance. It's a place that is silent except for the occasional birdsong, a place where dropping to your knees to sniff the ground might not seem so much odd as expected.

As a cool breeze wisps, and the bright sun shines on a spring day, the contemplation of just how perfect farm life could be sets in. Beyond the aesthetic is the idea of nurturing life—crops—and subsisting off of them. It's imagining going back to basics. Living simply. Then a turn 'round a bend changes everything. The future is there.

Apple corporation's enormous new data center is adjacent. It's 1.8 million square feet of glass and steel and new parts. There couldn't be a more stark difference in look or purpose. It's a mechanical beast producing not an oat or a carrot. It processes the intangible; these words (written on a Mac), it can be said.

The corporate takeover of land isn't exclusive of agriculture. Corporations have taken over our farms, too.

The average size farm in the United States, where large industrial farming thrives, is 443 acres. An average crop such as corn produces more than three tons per acre per season. The ground isn't meant to produce so much. To help it cope, soil is prepared with the help of sensors. Satellites also come into play. Aerial photos and data inform what the soil needs and where the weeds are. Nutrient data can also be pinpointed. Fertilizer, pesticide, and water quantities are mapped. After the soil is prepped, seeds, often bioengineered to tolerate drought or resist pests, are planted using machines. Automated irrigation methods begin.

Water is a big component of crop growth. While irrigation began in 6000 B.C. using floodwaters, it wasn't until three thousand years later that King Menes of Egypt began the practice in earnest, using dams and canals. Irrigation today is done with more precision, down to the drop, in some instances, with sophisticated technologies employed to maximize growth.

Watered, fertilized, and monitored for apex form, more pesticides are often sprayed to ward off intrusions and crop diseases. At harvesttime, combines and other heavy machinery are used to reap the most from the land. After that, the ground is tilled for the next season. And so on . . . until the soil has no more life to give.

Farms, make no mistake, are factories. And factories are meant to produce at maximum efficiency. Smart soil management is a cog in that machine. And the farms of the future will leverage technology and artificial intelligence to achieve the utmost yields possible. Big data and deep learning can, managed correctly, allow soil to thrive. Risk of overexploitation and degradation can be assessed using computer models. It won't be farmers scratching their heads at unproductive fields anymore. It will be analysts scrutinizing data.

Nearly half of all the land on Earth is farmland. Only 1 percent is used for urban living. The remainder of habitable land is covered by forest, shrubs, or fresh water. Three hundred years ago, less than 10 percent of the land was used for food production. The enormous difference between less than 10 percent of all land and the 50 percent today is hard to fathom.

As we clear more land for farms, and fertile soil shrinks, it's imperative we conjure alternatives to growing food—reengineering the poor job we have done of engineering wild vegetation for our dietary benefit. If farmland did not exist, the world would look vastly different from above. There would be more than thirty-two million square miles of green forest and vegetation growing wild. For visual context, the United States is less than four million square miles in size.

Ole Green's business is built around the farm of the future. He is the chief executive of Agrointelli, a company that utilizes technology to make intelligent solutions in the field. (Hence, the name.) The company develops virtual reality—like weed maps, robots, and even a sort of GPS for plows. He is prepared to discuss the farm of the future and what it might look like.

After arriving at his office at the Agro Food Park on the outskirts of Aarhus, a charming city in Denmark, and shaking every one of his employees' hands and saying individual "good morning"s—yes, every single one of the twenty-two or so people spread throughout different offices—he breaks open a PowerPoint presentation. It isn't what might be expected.

Green looks like he is in his late thirties. He likely wouldn't be mistaken for anything other than a Dane: light brown hair and goatee. Powerfully strident and energetic, he is a nonstop fast talker. So when he pops open his laptop and the first image of the PowerPoint appears, the pause he takes is noted. He waits for a reaction.

The image displayed is vintage. It's a farmer holding on to a plow that is being pulled by a horse. "This is smart farming," he says. "The farm of the future is exactly this." He dissects what he means: The farmer's eyes are 3-D technology. Reactions to the horse's pull are control systems. The farmer's feel of the plow are kinetic sensors. The horse's speed is an indicator of force, and informs about soil texture. The horse walking is yield impact data. Manure is biofuel. Everything that is the farm of the future is right there.

Cut to the next PowerPoint slide, where auto-piloted tractors replace the horse. Satellite images show soil texture. Hardware and software replace every element of the old photo. Technology and innovation only serve to make farming more precise. Which is why the future of farming boils down to precision farming. "Robots are not for farming but for automating tasks," Green says. "Artificial intelligence is a major driver, where robots can hear, feel, and sense like a farmer's experience."

Picture this: satellites and drones scanning vast land areas. They map water tables, soil depth, and type. Data is fed into a computer model powered by artificial intelligence. Weather and climate trends are baked in. The area is cordoned off and mapped in 3-D. Remote-controlled machines clear fields. Crop rows are spaced exactly to scale to allow optimum yields. More drones are dispatched to seed and plant. Nanotechnology sensors send signals to computer models about growth rates and deficiencies. Irrigation is computed. Robots pick crops.

All of these advancements exist today. They are all aimed at hyperengineering land for our benefit. But no matter how far advanced the technology and the tasks that can be automated, there is no replacing, in sum, the soil. Like water or air, we can't manufacture it, or at least not in vast amounts. The choice is to reengineer what we have. Which is why Green and Olesen are working together on the SmartSOIL program to blend innovation with best-in-class practices.

Still, efficiency and soil regeneration come at a price: additives. For smart soil programs to truly work, land cannot be overly tilled. That means weeds will grow. Weeds strangle healthy crops and crimp yields. Hence the reason why weed-killing herbicides are frequently used in no-till agriculture programs. Herbicides are technically pesticides, and their use can be deadly. Various studies have linked different types of herbicides to birth defects, cancer, and deaths. Some herbicides are, of course, less toxic than others.

Lars Munkholm, a senior researcher at Aarhus University's Department of Agroecology, says soil structure is complex business and there is no easy answer to the weed and disease issue. Using more or less herbicide, for that matter, depends on a multitude of factors including soil depth and how compact the soil might be. "There are many conditions," he says, as he takes time out of a Sunday to meet at the research center offices to discuss soil's future.

The same blunt question is posed to him and Olesen: "Can we unfuck what we have fucked up when it comes to the soil?" They both have more or less the same answer: "We have to try and we have to try to educate." Getting hung up on the "franken" side of things—using synthetic pesticides or fertilizers—is to miss the point. It's whatever it takes to make soil healthy again—intelligently. Nature can produce its own pathogens, too. The battle isn't man versus nature. Winning the war is achieving balance from both.

But educating small farm holders about which herbicide is better than another or about how to change practices that have likely spanned generations is no easy quest. There are 570 million farms in the world, and most are small and family run.

Education is the major impediment to smart soil programs being adapted on a meaningful scale. We can yammer all we want about regeneration, but it is going to take a whole lot of education to make soil smart again. Think back to those rogue operators out in the middle of the Borneo rainforest. Who is going to educate them? How are they to know that the fires they light also scorch arable land and take away prospects of growing food they rely on for themselves, maybe their families? This is the quagmire: consideration.

It's worth repeating the alarming prospect that all the world's topsoil may be eradicated before the next century arrives unless the Earth is rehabilitated and land reengineered. Food may have to be sourced differently.

Vertical Farming

 The world's largest vertical farm is in Newark, New Jersey—not the first city that comes to mind when conjuring images of agriculture.

Vertical farming is a way of growing food that is exactly how it sounds: vertically rather than horizontally. Typically done indoors, stacked shelves or other means are utilized for seedbeds instead of rows. The idea is to make maximum use of space, which is why many vertical farms can be found in urban environments.

AeroFarms, the company behind the New Jersey facility, uses advanced technology to grow crops indoors without sunlight or soil.

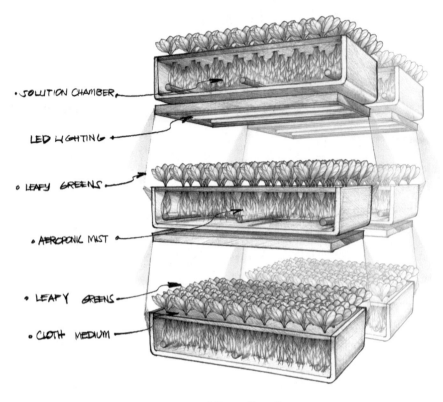

A Vertical Farm Crop Row

At the seventy-thousand-square-foot facility in Newark, Aero says it can harvest as much as two million pounds of leafy greens per year. The average yield for that crop from a traditional farm in the United States is a little more than thirty-six thousand pounds. Obviously, the difference in results is huge and is the reason why vertical farming is touted as the new "green revolution."

The company has patented technology that optimizes growing conditions and uses 95 percent less water than traditional field farming. And 40 percent less water than hydroponics, which is a common indoor technique. Growing plants by their roots in a water-nutrient solution has been long popularized by the cannabis, hothouse-flower, and other floral industries that require controlled environments. Artificial lights replace the need for sunlight, and plants can be stacked for maximum production—indoors and out of sight.

Many of us probably remember utilizing hydroponics in elementary school to learn about botany (likely with lettuce or basil roots, not cannabis). Aeroponics takes no-soil growing up a notch. AeroFarms says it uses progressive methods to achieve hyperyields: LED lights are used to create a specific light recipe for every plant. This provides the exact spectrum, intensity, and frequency needed for photosynthesis to occur in the most energy-efficient way possible. AeroFarms says it can take the same seed that might be used in a field and grow it in half the time. This leads to nearly four hundred times more productivity per square foot than a conventional outdoor farm.

Smart data is part of the magic. It is incorporated into the growing cycle. That means thousands of data points are analyzed to ensure consistent growth results. Indoor growing itself also helps with growth results. It lessens the chances for pest problems and diseases. AeroFarms takes disease prevention a step further and uses a special recyclable cloth for seeding and harvesting. The cloth leaves little chance of crop infection. The company says all these steps guarantee better yield per square foot.

Vertical farming is not new. Its basis dates back to greenhouses, which were common in Europe and Asia as far back as the thirteenth century. But the combination of technological advancements and urbanization brings about new possibilities. Vertical farming has caught on around the globe and in major cities. In London, a vertical farm operates underground in an old air-raid shelter. In Kyoto, Japan, farms are being taken indoors for the safer and higher-yielding environments vertical farms can offer. And some people much prefer food grown indoors. Celebrated chef David Chang is one of AeroFarm's partners. He says the greens have superior texture and flavor.

The catch with vertical farming is expense, as all that lighting and technology adds up. AeroFarms's Newark facility, for example, cost a reported thirty million dollars to build out. So much, in fact, that some observers question whether vertical farming can be profitable. Costs aside, growing more food on less space is what the world needs.

Drone Farmers

Pests and diseases can kill nearly 50 percent of crops in the developing world every year, and as much as a quarter of all crops in developed countries such as the United States. Manually scouting and then spraying pesticides has proven to be an imperfect way of warding off agricultural intruders. Enter drones. Equipped with infrared sensors, drones can scan vast areas of farmland and identify problem areas or even potential causes. Spraying can then be more refined.

AgEagle, a company based in Kansas, is one of a new batch of companies that has developed specialized drones that can scan fields and

The AgEagle RX-47

gather data. At AgEagle, the data is transmitted to FarmLens, a proprietary software platform that analyzes the data and produces a visual display of the area surveyed with detailed information about every square foot of cropland. Adjustments and solutions are recommended, such as more or less irrigation, or more or less agrochemical spraying coverage. Because the entire process is autonomous, predictive, and represents a farm as a series of data points, issues can be addressed instantly—and remotely. It's field information that the naked eye can't match.

Operating in about fifty countries, the company's agricultural drone looks like a mini stealth bomber and is equipped with multispectral sensors and specialized cameras. If this dystopian-looking machinery can significantly reduce the amount of crop losses globally, it can save more land from being sacrificed purely to feed us humans, meaning technology may come to nature's defense.

Small farms may also benefit from agtech. Millions of small family-run farms have gone bust over the past eighty years, since the advent of large, corporate farm culture. Corporate-run farms, running larger tracts of land, have adopted mass-food-production efficiencies that allow them to profit from scale. Small farms can't compete, and millions of acres of cropland have been abandoned. Advanced technologies can make small farms competitive. Robot pickers, soil-quality monitoring, big data, and drones, combined with artificial intelligence, may bring greater efficiencies back to smaller agricultural producers. That could make small farms feasible again and revive lost land.

CHAPTER 7

Brightening Clouds

There is nothing refreshing about swimming in the Persian Gulf. In fact, it actually feels cooler when you break through the surface after a dunk. The gulf regularly sees sea temperatures from 90°F to 95°F during summer, giving it and the Red Sea, just across the Arabian Peninsula, the ignoble distinction of being the world's warmest large bodies of water heated by the sun.

The air temperature may be 111°F, as it is on this day, but the small breeze that catches your exposed skin cools it when you come out of the sea. The pockets of heat beneath the surface, meanwhile, envelop your legs as you stand in the shallows. It's a weird, inverted experience for anyone used to taking a dip in most oceans or lakes. The average ocean temperature is about thirty degrees cooler.

Standing onshore, the seawater drips like sweat from your skin. It gets whisked by the wind, or runs fully down your body back into the gulf. Small waves gently wash onto the beach, quietly, nearly without sound. What you do hear are a lot of "ooh"s and "aah"s. The sounds aren't directed at the imperial blueness of the water, nor the beauty of the coastal Middle Eastern environs. They are the sounds of pain. The beach sand is scorching, and you run back to your towel, feet burning, like Dudley Moore in the movie *10*. Nearly everyone does it. "Ooh." "Aah."

As you dry, the salt stings. Hypersalinity has spiked over the past twenty years here, with saltwater concentration 1.5 times higher now. It's more skin-tingling evidence of climate change. Increased global temperatures induce more evaporation, leaving more salt in the water that remains. And some marine scientists claim the proliferation of desalination plants in the area is also exacerbating salt levels. Brine returned to the sea from the desalination process is dense with salt.

The marine life here suffers for it. The gulf may lose a significant portion of its biodiversity by the end of the century recent studies of the fisheries show. A scientific method called environmental niche modeling reveals that gulf areas suitable for marine life are shrinking and causing a loss of 12 percent of ecosystem biodiversity due to warmer global sea temperatures and higher salt levels.

More alarming: A fleet of Seaglider robots released into the adjacent Gulf of Oman—notorious for its piracy and political instability in places such as Pakistan—discovered a dead zone about the size of Florida just off the coast of Muscat and stretching far out into the Arabian Sea. Dead zones are where warm water and/or pollution diminish oxygen levels so much that it is difficult for animal life to survive.

Yet despite this deterioration in marine life, a curious phenomenon is happening just around the peninsula bend. The Red Sea, which also experiences almost the same exceedingly warm water temperatures as the Persian Gulf, is seeing its coral reefs thrive.

Most coral reefs around the world have fallen victim to global warming. The ocean's temperature change causes acidification that, in turn destroys coral. No such thing is occurring to the Red Sea coral, which makes it such an anomaly.

Scientists studying the corals say the six-thousand-year journey the organisms took to get to the area from the Indian Ocean, hundreds of miles away, has allowed them to acclimatize to higher temperatures. The shock of relatively quick temperature rise is what's killing off coral in other seas. Since 1901, sea surface temperatures have risen an average of 0.13°F per decade, and predictions are for them to continue to increase into the foreseeable future. One-tenth or so of a degree of temperature rise may not sound like much, but marine environments are super sensitive. Even such a small rise over decades can wipe clean life of all shapes and sizes and magnitudes, including coral.

Coral experts claim there is no other coral in the world experiencing the kind of resiliency being seen in the Red Sea because of that coral's unique journeyman characteristics, adjusting to temperature changes little by little over thousands of years.

Corals are technically living animals, as they do not participate in photosynthesis; rather, they rely on other sources for food. They feed upon zooplankton and other tiny, floating animals, and like all animals, they produce waste.

Corals, which we commonly just call "coral," are actually comprised of thousands of creatures called polyps. These are soft-bodied but grow a hard outer skeleton. The skeletons are what we often associate with coral. They attach themselves to seafloors, rocks, or other corals; bunch up into colonies; and come in fantastic displays of color. The algae that gets built up inside corals produces all sorts of different pigments and is what can be seen through the clear polyp bodies.

Corals are the biggest producers of life on Earth. Over the past twenty-five million years or so, they have cycled food into waste that algae photosynthesizes. This allows the coral colonies to grow and expand into reefs. The reefs then become habitats for larger forms of marine life, and the building blocks of the entire ocean ecosystem are created.

But corals are sensitive to changes in their environments. If temperatures rise too much or there is more sunlight, or if the nutrients they rely on diminish (such as with zooplankton—an occurrence explained in more detail in the next chapter), then corals discharge the algae that has grown inside them. This makes the coral turn white, or what is known as coral bleaching.

Bleaching brings corals to the brink of death and can eventually kill them. Globally, coral bleaching is happening at unprecedented levels, at least since record-keeping began in the 1880s. Seventy percent of the world's coral reefs are exposed to higher ocean temperatures, threatening their existence. From there it's a skip, not a hop, that threatens humans, too. We depend on the reefs, from the seafood they produce, which is the single most important source of protein for humans, to the oxygen, which corals help generate. The world's reefs are important to preserve.

The country of Jordan is taking the radical step of moving its dying coral reefs to the Red Sea to try to help rejuvenate them. By artificially integrating them with the ancient and resilient corals there, the hope is that the replanted reefs will survive.

"In 2012, corals from the southern region of the coast and the Al Derreh area were placed in baskets by a team of divers working on the project and transported almost two miles north, while continuously submerged underwater. Using marine cement and metal structures, the latter of which was created solely for the translocated coral, the corals were then replanted at damaged reefs and a cave site. Smaller coral colonies were moved to a nursery site. After a protection period to ensure the transplants' success, the new sites, just in front of the Aqaba Marine Park, opened to the public in 2018," *National Geographic* reported.

If the reengineering process proves successful over time (it's being closely monitored), other reefs may initiate similar "reseeding." It could be yet another trial of hope for the Great Barrier Reef, off the coast of Australia. It is the world's largest reef, and the largest living thing on Earth, measuring at about the size of Italy. Its enormity can be easily spotted from outer space. But what could be a spectacular display of life is merely a reminder of death and our unruly planet.

There are lots of plans to try to save the Great Barrier Reef from the numerous threats responsible for its decline—climate change being the most significant. There's a plan to pay Australian farmers to pollute less so toxic runoff doesn't end up in the ocean and destroy the coral. When farmers use harmful synthetic fertilizers, the wash ends up in streams and rivers that make their way out to sea, eventually landing on the reef and infecting it. There's another plan to unleash into the depths giant sea snails that eat starfish, which also play a part in destroying the reef. Crown-of-thorns starfish eat coral, and there have been outbreaks of the spiny creatures of late, jeopardizing big sections of the reef. The reason for the starfish outbreaks is unknown, but speculation is it's due to climate change. There are yet other plans to increase reef patrols to police illegal overfishing, which ruins the ecosystem, and in turn, the coral population. Without big fish to eat small fish, and small fish to eat smaller organisms, algae can build up on coral and suffocate it. Surfing robots have even been dispatched to monitor the Australian coastline to collect important data to assist with better cleanup efforts. The robots are called Wave Gliders and autonomously pilot around, gathering weather and water-quality measurements.

But no plan to save coral is more ambitious than the one John Latham, a British-born cloud physicist, and Stephen Salter, a South African–born design engineer, dreamed up: They aim to brighten the sky in order to cool ocean waters and save what lies beneath. Their aim wasn't initially directed at the Great Barrier Reef, or any reef specifically, for that matter. But it turns out the reef and other endangered marine habitats could be the biggest beneficiaries of their bold idea to thwart global warming.

Latham is a revered British physicist who has researched cloud formations for nearly half a century. He's won numerous meteorological awards and founded the University of Manchester's Centre for Atmospheric Science. He is also a prizewinning poet, which provides some insight into his creative approach to climate science.

Salter, too, approaches scientific problems in different ways. At eighty-one, he is a hoot to speak with, and his tenure teaching at the University of Edinburgh comes through in his ease of translating the academic into the practical.

Latham's idea was to brighten marine clouds to better reflect the sun's energy, hence cooling the water below. Salter designed an audacious yet practical way to do it: An armada of auto-piloted yachts would traverse the oceans and spray seawater into the air to brighten clouds and better reflect the sun. The albedo effect, as we've learned, is how much of the sun's energy gets reflected back into space by means of white or lighter-colored surfaces. Clouds do this naturally. But making them brighter by spraying seawater into their cores could increase their albedo capabilities and cool things even more—enough to balance the Earth's temperature even if carbon dioxide emissions doubled, according to Latham's climate models.

The marine cloud brightening idea stems all the way back to 1990 when Latham penned an article for the journal *Nature* titled "Control of Global Warming?" As a cloud physicist, he knew the power of clouds to manipulate temperatures as well as weather. And he also knew what kinds of things altered the composition of clouds. Ships crossing the oceans, for example, brighten clouds over water. The sulfate from their engine emissions wafts up and mixes with the gases that form marine clouds. They become markedly brighter, as satellite images prove.

Marine clouds are different than clouds over land. They usually hover much lower in the sky and reflect about 10 percent less sunlight. Marine clouds carry bigger-sized water droplets than continental clouds, as well. Smaller water droplets are more reflective because of something called the Twomey effect, which explains that more concentrated amounts of water droplets form a virtual shield and shine bright. The effect is facilitated by aerosols or pollution. Ships prove this when their emissions are added to the mix of maritime clouds. The clouds become more reflective.

Accompanying Latham's original article on the subject of marine cloud brightening is an image of ship tracks on the Atlantic Ocean in the Bay of Biscay just off the coast of France. It clearly shows brighter clouds above where ships had just crossed when compared with areas where there were no ship tracks. A Rorschach-like image, Polaroid-sized, reveals oblong patches of black—the sea—surrounded by thin gray stretches that layer in mishmash fashion—stratus clouds. There are hatch marks next to them: ship tracks. Their patterns are linear, unnatural on the dark and choppy backdrop. Above them, looming larger in size for the perception of depth, are fuzzy white blotches, their fringes sparked bright, like fireworks. These are the clouds that have been conditioned by ships.

Latham's big idea was to replicate this phenomenon, replacing the sulfur emission of the ship's engines with saltwater, which similarly brings about the Twomey effect. On a huge scale, the result would be to effect global cooling. But in addition to global cooling, Latham discovered marine cloud brightening can accomplish more localized tasks, too, such as preserving coral reefs, averting the collapse of ice sheets, or reducing the strength of hurricanes.

"[Marine cloud brightening] differs from some other [solar radiation management] strategies in that it could in principle also be deployed on a much less than global scale, thus offering a range of regional applications. On the one hand, it could possibly produce a localized ocean surface cooling, which could be applied to mitigate, or even avert, certain regional consequences of warming, which are being observed now, or which are predicted to occur in coming decades," Latham wrote.

After his article was published, he approached Salter about teaming up on a project to bring his marine cloud brightening thesis to life.

Salter invented Salter's Duck, a device that bobs (like a duck) in the ocean and converts wave power to electricity. He was intrigued by Latham's idea. He liked the concept of saving the planet, of course. He also liked the challenge of how to make a self-sustaining device that could distribute precise amounts of seawater into the air and have it attached to a vessel that could indefinitely roam the seas without the need for human piloting.

What he conjured up was this: a 130-foot, 300-ton catamaran and apparatus that look like giant stovepipes. Renderings show a sleek white yacht that could just as easily be billed as a Caribbean leisure-cruise charter as an advanced scientific research vessel carrying hope for a cooler blue planet.

Salter said the first thing he had to think about was energy. He'd need two sources of power: one to move the vessel and the other for the device, to turn the rotors that would push air and seawater up through a large spout. Tracking devices and computers would also need energy to operate.

For motion, he decided on wind power: a sailboat that skims along the surface of the water much like an America's Cup yacht. Hyperlight foils allow these superyachts to glide along with less drag, like hovercrafts. Salter similarly needed as little drag, or friction, as possible. He wasn't concerned about speed. He needed to combat the drag associated with funneling water for the salt spray.

The Albedo Yacht

Instead of traditional sails, Salter turned to Flettner rotors, which are vertical cylinders that capture wind in discs and propel the vessel forward. The next thing he had to figure out was filtration. What type of system could be engineered that wouldn't clog? Once the water was scooped from the ocean, it needed to make its way through the rotors that, according to Salter's design, rise sixty-five feet high. From there, saltwater would spray out the top. A clog would shut the whole process down. He ended up turning to the membranes used to filter out the microscopic polio virus from water. Polio can be transmitted by water, and effective filters have been developed to purge it. (Although vaccines should be credited with largely eradicating the disease.) In any case, polio filtration holes are nanometers small and aren't prone to clogging. A billion and a half of them punctured through a silicon wafer twenty centimeters in size did the trick for Salter. With that, he could produce the jet stream he needed to create droplets of the particular size that would best interact with cloud condensation.

In Salter's design, once the vessel is moving, water is dragged through turbines and forced through the rotors and pipes that funnel up and out the top like a showerhead. The size of the drops produced (0.8 micron) is important. They have to be small enough so they can waft up high and make it into clouds. Big drops fall back down into the ocean too quickly.

Once the drops coalesce with the cloud's condensation, the clouds brighten. The brighter shield better deflects the sun's radiation. In scientific terminology this is called radiative forcing. Marine cloud brightening fosters negative radiative forcing, which means it has the ability to stop local and, in turn, global temperatures from rising.

At the top of the Earth's atmosphere, the sun hits areas facing it directly with 1,360 watts of energy every square meter. Spread that energy over the size of the planet receiving sunlight and it's a terrific amount—enough in mere minutes to power the entire global population's energy needs for a year. More than half of that energy gets absorbed into the atmosphere, mostly by clouds, before making its way to the Earth's surface. Brightening clouds with seawater spray dims the amount of energy reaching the ground, or in the case of the Great Barrier Reef, the ocean.

It seems rather innocuous, marine cloud brightening. Nothing artificial is added to the atmosphere. The only thing the albedo yachts, as they've come to be known, do is facilitate ocean spray and loft it high into the air. But it turns out there may be consequences, big consequences, that could bring about unexpected changes to weather patterns.

Teleconnection patterns are large-scale meteorological impacts. They can be likened to the theoretical butterfly effect. They explain how and why weather in one place affects weather a continent or an ocean away. The patterns string out around the globe and allow meteorologists to track changes. El Niño is an often discussed teleconnection pattern, as is its opposite, La Niña.

The Climate Prediction Center of the United States National Weather Service defines these patterns as changes in the atmospheric wave and jet streams. The changes influence temperatures, rainfall, and storm intensities.

El Niño is technically the Southern Oscillation pattern and means abnormally warm water has developed off the Pacific Coast of South America. That brings more rainfall to places such as the US Gulf Coast and generally milder winters in the northeast United States. La Niña typically brings about the opposite effect and is caused by cooler waters in the same area off the coast of Peru. El Niño loosely means "Christ child" in Spanish. It was coined two centuries ago by Peruvian fishermen who noticed the warmer waters around Christmastime.

Interrupt these age-old patterns with synthetic ocean cooling, as marine cloud brightening would bring about, and weather might spin unpredictably out of control. Some climatologists believe that if marine cloud brightening were to be conducted in the Atlantic Ocean, for example, it could lead to the Amazon rainforest completely drying up, which would be a catastrophic result: The Amazon provides the world with vast amounts of oxygen. Moreover, "while climate responses to geo-engineering have been studied in detail, the potential biodiversity conse-quences are largely unknown. To avoid extinction, species must either adapt or move to track shifting climates," argues Christopher Trisos, a researcher at the University of Maryland.

Salter noted that the article did not mention marine cloud brighten-ing by name but was "hostile to all kinds of geoengineering." He took issue with its results and says marine cloud brightening can be calibrated so it would do no such harm to the planet. Of course he also promoted the idea of more testing and experiments.

Trisos expanded his critique in a conversation about marine cloud brightening and said sudden implementation or, for that matter, sudden termination—a phenomenon known as termination shock—could be devastating to species and be like nothing the planet has ever experi-enced. Animals and plants, even us humans, adapt to the particulars of a climate. And some species can survive only in certain conditions. Tropical plants, for example, die quickly when exposed to colder tem-peratures. Extrapolate that scenario across all species on the planet and a massive termination effect plays out.

"Because we depend on ecosystems for our survival, we should un-derstand how solar radiation management affects them," Trisos said, noting that when he first learned of solar geoengineering he thought that it was crazy and something of science fiction. Then he realized that there are serious programs being readied for testing. That's when he dug into his research. Most of his study has been on stratospheric aerosol injection, where clouds are seeded with particles to increase re-flection. But he's also looked at marine cloud brightening. "Both have a termination risk," he said. Suddenly turn off the experiment, and tem-peratures shoot back up. Going back to his adaptation concerns, there is scant time for species to adjust to the spike. With stratospheric aerosol injection, there is a lag of several months before temperatures creep back up. That means if the proverbial wheels fall off a program, there is time to reboot the project, or figure another solution. But with marine cloud brightening, "you would lose your cloud cover very rapidly . . . within days," he said. "And if you already had a warmer planet, that would be a really bad thing." Besides droughts, large-scale wildfires could erupt, Trisos adds.

He emphasizes that this is all theoretical. There has been very little research done on the effects of solar radiation management, and even less investigation into the effects of marine cloud brightening. It's cause for concern.

Still, scientists in Australia at the Sydney Institute of Marine Science believe marine cloud brightening has the best chance of saving the Great Barrier Reef.

Daniel Harrison, a researcher at the University of Sydney, said he and a team of other scientists examined all sorts of approaches to saving the reef, and marine cloud brightening holds the most promise because it can be targeted. Small sections of the reef can be tested, and if the process works, "a broad scale process could scale up for the whole reef," he says.

Comprehensive modeling has to be done to ensure the ocean area would cool with maximum benefit to the corals, but he is encouraged with model results to date. The Australian government is now funding a full-on feasibility study of marine cloud brightening as a reef-saving method.

While vessels along the lines of what Salter designed could be used for marine cloud brightening above the Great Barrier Reef, Harrison said there are many options: land-based spray devices; snowblower-like machines that would be placed on ships; or floating platforms onto which stationary devices could be affixed.

In terms of negative consequences, Harrison isn't too worried. "The effect would be very minimal," he said. The fact that marine cloud brightening would take place on a localized basis would mitigate the possibilities of larger tagalong effects, he says. That said, he's planning to incorporate atmospheric models in the ocean, water-quality and reef analysis models that he and his team are building.

If indeed harmful pass-along results occur within the computer models, the experiment will be shut down. "It might not work at all. Or it might work better than we thought," Harrison said.

For the purposes of cooling the Great Barrier Reef, marine cloud brightening would only be utilized intermittently. Harrison said that would mean spraying for just a couple of weeks at a time in coral areas that are most in danger. And the spraying would likely take place only in summer, when ocean waters around the reef are warmest.

He is forthright about the fact that there is no silver bullet for lowering ocean temperatures in general and, in turn, the reef. "It doesn't negate the need for mitigating carbon emissions," he says.

Of course, dialing back carbon emissions isn't happening as quickly or as effectively as once hoped. The oceans are still on track for unprecedented warming over the coming decades.

Could a fleet of albedo yachts dispatched to ports around the world help save the day? It's not an easy or inexpensive proposition, never mind the possible negative consequences.

Salter estimates a fleet of fifty vessels launched annually and costing around two million dollars each could negate a year's worth of global carbon dioxide emissions. It could hold ocean temperatures in check. Compared to the economic losses linked to global warming—trillions of dollars—marine cloud brightening is a cheap fix.

A group of Silicon Valley investors almost pulled the trigger to fund the construction of the albedo yachts for a pilot program. But that scheme fell apart. The UK government also floated the idea of funding the project, but that never transpired, either.

Latham and Salter are still hopeful, however, that funding will come about and their big dream will come true.

Ultimately, they foresee fifteen hundred auto-piloted yachts traversing the seas and spraying seawater into the air like a spouting pod of humpback whales. The yachts would be directed to hot spots by season and by remote control. And they could be diverted from places where bad weather would impair sailing conditions. Areas such as the Great Barrier Reef would get treatment as needed, and then the fleet would move on—constantly seeking out brightness, showering the sky and cooling waters the world over. If you can imagine seeing a rainbow through that mist, well, that's the dream.

Sunlight and Seaweed

Seaweed is often called the oceans' "trees." And like trees, seaweed sequesters carbon out of the atmosphere and shades places. In the oceans, lower carbon content can mean cooler water temperatures and lower levels of acidity. Acidity occurs when ocean temperatures heat up, destroying marine life and habitats such as coral reefs.

So why not grow more seaweed to protect oceans and ocean life? That's what Tim Flannery, a renowned climate scientist and author of *Sunlight and Seaweed: An Argument for How to Feed, Power and Clean Up the World*, believes we should do. His plan? To geoengineer massive seaweed farms around the world.

Flannery is a leading voice in Australia on climate issues, and was for twelve years the country's climate commissioner. He is likely best known for another book of his, *The Weather Makers*, in which he chronicles how humans have been changing the climate and what it means for life on Earth. He also predicts how potentially cataclysmic climate events will unfold over the coming century. An explorer and conservationist (he often affects that Indiana Jones look in hat and bush shirt), he grounds his solutions in a combination of academic research and fieldwork. He is a highly respected member of the scientific community. Which is why his seaweed solution is worth serious attention.

Seaweed is kelp. It grows quickly and abundantly and is also a major source of protein for billions of people. Likewise, seaweed is essential for sea life. Storing carbon, promoting biodiversity, and providing thousands of species with places to live and grow, seaweed serves as the foundation for much of our marine ecosystems.

Flannery's plan riffs off analysis by researchers at the University of the South Pacific who modeled out what the world would look like if nine percent of the ocean were covered by seaweed. The outcome was, as Flannery put it, "stupendous." It could negate the amount of carbon emissions humans put into the atmosphere, create a new energy source (biodigested methane), and foster food for livestock and humans alike.

There are many different types of seaweed. Some humans eat. And some kelp is used as animal feed on farms. Kelp is a multibillion-dollar, global business.

"Seaweeds can grow very fast—at rates more than 30 times that of land-based plants. Because they de-acidify seawater, making it easier for anything with a shell to grow, they are also the key to shellfish production. And by drawing CO_2 out of the ocean waters (thereby allowing the oceans to absorb more CO_2 from the atmosphere) they help fight climate change," Flannery says in his book, which reads like a manifesto.

Growing kelp is done all the time close to shore and also on land where it can be managed effectively. What Flannery proposes is offshore kelp farming, far out at sea. Mid-ocean kelp farming has also been tried. Bad weather and poor materials and designs caused the farms to fail miserably, however, despite billions of dollars of investments. With better siting, more advanced technologies, and more resilient materials, Flannery believes kelp farming can work. He cites a sustainable design put forward by Dr. Brian Von Herzen of the Climate Foundation: "a frame structure, most likely composed of a carbon polymer, up to a square kilometer in extent and sunk far enough below the surface (about 25 meters) to avoid being a shipping hazard. Planted with kelp, the frame would be interspersed with containers for shellfish and other kinds of fish as well. There would be no netting, but a kind of free-range aquaculture based on providing habitat to keep fish on location. Robotic removal of encrusting organisms would probably also be part of the facility. The marine permaculture would be designed to clip the bottom of the waves during heavy seas. Below it, a pipe reaching down to 200–500 meters would bring cool, nutrient-rich water to the frame, where it would be reticulated over the growing kelp." The system would be powered by solar energy. Docks and refrigeration systems could also be attached to the floating kelp frame. There are myriad iterations to the kelp-farming solution. They each involve various ways to grow and harvest the ocean's most abundant crop. With more farms and more crops also come more shade. A brighter future for the oceans means we may have to darken the waters with seaweed to cool temperatures and keep marine life alive.

Spraying the Stratosphere

He may be the closest thing in the world to a Dr. Frankenstein for the climate. An intense bespectacled and bearded research scientist in his late fifties, he looks the part, too. This is David Keith, somewhat of a living legend in geoengineering circles. The Harvard University professor and climate scientist is an outspoken proponent of climate intervention. He appears on television talk shows in snappy suits and articulately lays out his case for artificially altering the atmosphere. At lectures, he defends his position by noting that some people—even many people—could die as a result of his solar engineering: spraying aerosols into the stratosphere to redirect sunlight.

The stratosphere begins about twelve miles up, and it is the next layer of the atmosphere from where we live on the Earth's surface. The stratosphere is also where the ozone layer is located.

Keith's plan is to inject different types of materials into the stratosphere to figure out what works best to reflect the sun's energy and diffuse it. The plan, Stratospheric Controlled Perturbation Experiment (SCoPEx), is a bold test. A balloon will be launched a little more than twelve miles up into the atmosphere. It will carry a bevy of instruments and as much as two pounds, or a little more, of aerosol materials that will be released into the sky. The air mass at that altitude will stretch more than a half-mile long and be several hundred feet wide. The instruments on board will then measure how dense the mass becomes, how the aerosols interact with other matter, and how light scatters.

Keith says he plans to experiment with different types of materials such as ice and sulfur to measure results. Whatever is used won't much disturb air patterns. According to the project description maintained by Harvard, "if we test sulfate in this experiment, the amount we would use would be less than the amount released during one minute of flight of a typical commercial aircraft. Aircraft release sulfates due to residual sulfur content of aviation fuel."

The hope is that SCoPEx will inform climate models and indicate how large-scale solar engineering projects would affect the planet—the kind that could come with deadly consequences

If the initial experiments, which are to be conducted over the coming years, prove positive, larger tests could follow. Balloons likely will be replaced by aircraft to spray larger areas. That's when major effects would then kick in. Every living thing on the planet would be affected. Ocean temperatures and our average global temperature would, ideally, be reduced. However, not all areas would be affected in the same way. The temperatures at the poles would be influenced differently than regions closer to the equator, for example. Hence, SCoPEx's aim is to figure out what those possibilities and differences are on both land and sea.

Stratospheric aerosol injections, such as the ones with which Keith is experimenting, are radical forms of geoengineering. Effectively, the process involves spraying pollution into the air, as that is what sulfur is at its core. Unlike marine cloud brightening, the aerosols are injected sky high, rather than sea-surface low.

The health effects from stratospheric injections of pollutants, of course, could be palpable. There could be ozone depletion. The ozone layer is known as the Earth's "sunscreen" and protects us from too much ultraviolet light. And stratospheric aerosol injection might reduce rain and snow, precipitation. Biodiversity would possibly be corrupted, and more unknown consequences could ensue. Again, SCoPEx will be used to figure out these ramifications.

The side effects would likely only occur if millions of tons of aerosols were sprayed. And Keith, who speaks of the downside with candor, soberingly emphasizes the upside, that we would be able to turn down the Earth's temperature (by how much is yet to be determined) and reduce the chance of a doomsday scenario.

SCoPEx, then, is more than an experiment: It's a preview for the balancing act we are going to have to face in the future.

CHAPTER 8

Fertilizing Oceans

Reeds bend in the light breeze. Ripples on the marsh water scramble the shadows of cypress trees and water lilies. The dark images of a flock of herons come and go. Pockets of muddy banks weave in soft turns out of sight, where the last parcels of solid earth give way and disappear beneath the wash of gulf water. The natural beauty is a by-product of all that exists within wetlands—biological supersystems—that can hold more life than any other type of land or sea area, including rainforests and coral. So it's a sad truth that fish arrive here and die.

The Gulf of Mexico is one of the largest ocean dead zones ever measured. An area the size of New Jersey sits just off the Gulf Coast and deprives fish of oxygen. It petrifies the ecosystem and sets off a litany of effects that crimp food supplies, harm fishermen, and can kill, serially. And it's not nature's doing; it's ours.

Nutrient pollution from the mighty Mississippi River dumps into gulf waters and produces a toxic bath for fish and the coastal marine environment. Pollutants from synthetic fertilizers used by farms in the Midwest, sewage water, and other gnarly runoff get mixed into the Mississippi and work their way down to the Gulf. There the toxins form into algae ponds that hoard vast amounts of oxygen, depriving fish and the rest of the marine environment of it. Fish, just like us humans, need oxygen to survive.

Gulf waters provide the United States with 40 percent of its fish. The effect of the dead zone on food supply in the United States, therefore, is considerable. Employment, too, is affected. Less fish, less money for fishermen. The coastal economies of Texas, Louisiana, and Florida fluctuate based on fish hauls.

When BP's Deepwater Horizon accident in 2010 sent millions of gallons of oil spilling into the gulf, for example, it forced the closure of more than a third of the fishing waters. The regional economy tanked, and thousands of people were left unemployed. It still hasn't fully recovered. And that was just one accident.

The gulf's dead zone is nearly nine thousand square miles and has been growing for years. It's not coming back to life anytime soon. It can take a millennia for oxygen-deprived zones in the ocean to rejuvenate naturally. Which is why scientists are intervening to figure out ways to bring the ocean back to life.

The National Oceanic and Atmospheric Administration (NOAA) is encouraging greenstick fishing in the Gulf. Greensticking is a technique being promoted as an alternative to long-line commercial fishing, where unintended catch can get caught in the lines, including threatened species. It involves installing a tall fiberglass pole (originally these were green in color, hence the name) that stands high above the boat. From it, a long hook line is towed. Squid lures are attached, but they skip above the water, not sink below the surface. This attracts yellowfin tuna, which have the ability to scan the air above the water's surface for prey. Fish that don't have this ability or strength are saved from getting caught. And those that are caught unintentionally can be quickly released.

Traditional long-line fishing kills whatever is caught in its hooks. It utilizes a main line that can stretch for miles and to which thousands of hooks are attached. The line is submerged deep into the ocean and can, as might be expected, snag all sorts of indiscriminate marine life.

Fishermen specialize. They go for tuna or snapper or swordfish or what have you. They get licenses and quotas for their hauls. What they can't take in, they toss back into the sea. Long-line hauling produces tons of disallowed fish. That takes away from the possibility of bringing more of what is allowed onboard. The hope is that by concentrating fishing techniques, catches will increase. And while the evidence is still anecdotal, individual studies show that by greensticking, catches of the wanted species of fish increase exponentially—in some cases by eighty percent.

By increasing fish stock and better preserving marine ecosystems, the sea's natural defenses against dead zones—more life—can be emboldened. Marine ecosystems support life and healthy environments by their food chains and the waste produced from it. Fish waste, for example, supplies nutrients to smaller organisms that, of course, feed larger organisms, and so on. This naturally filters algae and creates oxygenized habitats that combat dead zone intrusions.

A NOAA researcher in New Orleans who was on her way to the gulf to document greensticking procedures and record catch numbers, says overfishing is a big, big problem, exacerbated by the dead zone and the Deepwater incident. "It's bad," she said. "There aren't enough fish." Bluefin tuna are in especially low numbers. The hope, she said, is to get fishermen to change their tactics and, in turn, help depleted fish populations to grow.

About two hours south of the airport is the southernmost point of Louisiana and where the last streams of the Mississippi River meet the Gulf of Mexico. The famous bayous plain out for miles in all directions. Roadways run level with the water's surface. Oil refineries such as Halliburton and ConocoPhillips litter this last section of the Mississippi. The refineries are, of course, ugly structures of steel and concrete that appear against the backdrop of otherwise unspoiled nature. Sprawling wide for acres and tall for 150 feet, they make an indelible impression. They are, in a strange way, intimidating. They exude their price tag—hundreds of millions of dollars—and purpose—brazenly constructed for facilitating industry. They are the most unnatural behemoths one can imagine.

A fire is burning atop one refinery's tower. It's the flare stack, a defense mechanism found at all petroleum refineries in case there is an emergency and the hydrocarbons need to get burned into carbon dioxide before escaping into the atmosphere. An unlit gas cloud could be far more hazardous because unadulterated gases would be released into the air rather than getting burned off. So the torch is always kept.

About a mile away from that flare stack, in the little harbor town of Venice, two representatives from the Louisiana Department of Wildlife and Fisheries (LDWF) were examining fish at the dock. They said they were measuring fish bones, heads, and tails to figure how many fish, and what type, are being caught so as to better regulate quotas.

Quotas limit the amount of fish such as snapper, grouper, tuna, and swordfish that can be caught by commercial anglers. Limits are needed so fish populations can be maintained. Otherwise, there are too few fish left to spawn and species begin to die off.

Sitting at the marina bar at Crawgator's Bar & Grill with a po' boy sandwich and a beer and listening to country music lyrics—"She's a flat-liner," "He can't even bait a hook," "Dirt road diary"—one watches the quotidian harmony of the fishing industry play out on a warm, sunny day. Boats head out to sea and return to docks. Catches are unloaded and brought to the processing center. Total amounts are tallied. Nature's bounty is converted into capital. Day after day the scene repeats. All over the world this scenario takes place in its local context. The bigger problems, the shocking facts, aren't evident.

Ninety percent of global fish stocks in the oceans have "alarmingly" low numbers of fish, according to the Food and Agriculture Organization. Overfishing and warmer waters attributed to climate change are cited as the main reasons for the diminishing populations. An increasing global population of humans coupled with more people eating fish as part of their daily diet is causing overfishing to meet higher demands. As previously mentioned, about a billion people around the world rely on fish as their major source of daily protein. And warmer ocean temperatures hinder the ability for fish to reproduce. Sea surface temperatures have risen every decade since 1900, and over the last three decades temperatures have risen more than at any time since readings began. Expectations are for continued rising into the foreseeable future.

A comprehensive study of marine life by a group of ecologists and economists speculates that all of the world's fish will be gone by 2050 if consumption and marine environment corruption continue at the present rate. The number of farmed fish consumed from the ocean has recently overtaken the number of wild fish consumed. Put another way, that means the human engineering of marine environments now produces more fish than natural spawning. But fishing innovations alone won't be enough to solve the ocean's problems.

There are hundreds of dead zones around the world, increasing in total size by millions of square miles since the 1950s.

It can take oceans a thousand years to recover from depleted oxygen levels, such as those in the Gulf of Mexico, according to one study by researchers at the University of California, Davis. They examined seafloor creatures from the last glacial period, which ended some twelve thousand years ago, and discovered that climate change disturbs seafloor ecosystems so extensively that it takes them a millennia to return to life. Hypoxia, or lack of oxygen, was the problem then as is the problem now; increasingly so. Researchers have found that the oceans have lost 2 percent of their oxygen over the last fifty years and may lose as much as 7 percent over the next fifty. Even a small decrease in oxygen levels can be catastrophic for marine habitats.

Just as on land, plants are largely responsible for producing oxygen. Ocean plants, such as kelp, phytoplankton, and algae, conduct photosynthesis—the process of converting sunlight and carbon dioxide into oxygen—for the benefit of sea life, and humans, too. In fact, about half the oxygen in the air is produced by marine plants.

The progenitors of oxygen in the seas are tiny organisms—plankton. Plankton comes in different forms, and each organism reacts differently to different types of marine environments. Warm, cold, shallow, and deep water, as well as local ecosystems, produce a plethora of variables.

Phytoplankton, which are small, single-cell organisms that float near the surface, are important to ocean life as well as life on land. They are the first link in the food chain for marine life. When phytoplankton die, they drift to the bottom of the seafloor and create a massive carbon storage layer. If phytoplankton are eaten, the carbon is obviously transferred to the next food chain link and is processed through that organism's waste or when it dies. All that carbon stored on the seafloor is the reason oceans are critical to managing Earth's temperature. The less carbon stored in the seas, the more left above the surface to absorb heat, and the warmer the world gets. It's a balancing act.

Zooplankton feed on carbon-capturing phytoplankton. Corals, as detailed in the previous chapter, and many species of fish feed on zooplankton. And bigger fish eat those fish. And so on. But the phytoplankton are dying. As much as 40 percent of the phytoplankton in the seas have died since 1950, according to marine studies. If that year rings familiar, it should; it brings us back to the proliferation of dead zones. Pollution and warmer temperatures, among other factors, are causing the huge decline. Given that phytoplankton are so low down on the marine life food chain, all the life above it suffers, too. Fish of all sizes are taken out of the food supply chain. Which brings up the importance of whale poop and the Sahara Desert. No need to read that twice. The curious connection between whale feces and desert sands makes sense when you learn that both contain high concentrations of iron. And phytoplankton need iron to grow. African dust storms from the Sahara dump large amounts of sand into the Atlantic, helping phytoplankton to grow there. Dust also makes its way to the Amazon to feed plants on land, as well.

Whales, before they were hunted and slaughtered en masse, apparently also helped phytoplankton grow and thrive. By one account, their waste was responsible for twelve percent of the iron on the surface of the Southern Ocean.

Human interference has dampened phytoplankton's ability to grow naturally, though. Clearly, commercial whaling isn't the sole cause of the interruption. The whaling industry's heyday was in the mid-1800s, after all. Rather, profligate plastic pollution, nitrogen pollution, and warmer waters are dimming phytoplankton survival.

There are four major oxygen-deficient zones in the world: the eastern tropical North Pacific, off the Guatemalan coast; the South Pacific, nearer to Australia; the Bay of Bengal in the Indian Ocean; and the one in the Arabian Sea that we already discussed. Now there are other unexpected zones, such as in the Gulf of Mexico, and in the northeast Pacific's Subarctic Oceans, also.

The northeast Pacific dead zone shouldn't logically be there because it experiences lots of upwelling, creating nutrients. Micronutrients from the bottom of the ocean cycle up to the surface and feed phytoplankton at the top. (Upwelling also creates fog, which is why it is so misty in the Pacific Northwest in summer.)

The aberrant dead zone in the northeast Pacific really bothered John Martin, a scientist at Moss Landing Marine Laboratories in California. So much so, he decided in the late 1980s to test the waters. He found there was a serious iron deficiency. Technically this is defined as a high-nutrient, low-chlorophyll (HNLC) zone. Like any good doctor might prescribe when a person has an iron deficiency, Martin suggested adding more iron to the ocean's diet. He conducted a small experiment to test the results.

In a January 1988 article for the journal *Nature*, Martin said phytoplankton increased in proportion to the amount of iron added. The experiment wasn't enough to prove his hypothesis as an ocean cure, but it was a beginning. Yet the marine science community begged him to stop.

The prospect of "ocean engineering" with iron additives appalled them. Martin was harshly criticized for wanting to litter the oceans with iron dust. The criticism, however, only egged him on.

"John Martin planned on following up these initial tests by fertilizing a small patch of [HNLC] ocean in the central pacific near the Galápagos Islands. If the phytoplankton populations shot up with the addition of iron, then his hypothesis would be proven correct. In 1991, however, Martin began having back pains. Medical tests revealed he had prostate cancer that had spread into other parts of his body. For the next two years he underwent chemotherapy and radiation treatments," wrote NASA's Earth Observatory in a tribute to him. Martin died in 1993. His experiment carried on. Other scientists from his organization completed the Galápagos tests that same year and found phytoplankton levels zoomed.

Still, the marine science community didn't afford the tests much due. The reason has to do with complexities. Another experiment conducted by one of Martin's colleagues showed certain types of phytoplankton reacted differently than the initial tests indicated. Not as much carbon was absorbed and fewer plankton were produced. In fact, even further tests revealed that dangerous forms of plankton could be produced by iron dusting, the types of plankton that cause "red tides," or algal bloom that produce toxins perilous to fish and can make people sick if they eat shellfish.

But some individuals were struck by the positive possibilities of Martin's hypothesis. Entrepreneurs saw profit potential from iron dusting. Every ton of iron added to the ocean could remove as much as 110,000 tons of carbon from the atmosphere, Martin's tests showed. By creating carbon "offsets" or trading schemes around successful ocean fertilization, big profits could be made.

Carbon offsets are "credits" polluters can buy to negate their emissions. The credits can be purchased from, in a simplified example, an organization that plants trees. (Thereby storing carbon.) Credits can also be traded. The scheme encourages environmentally friendly practices while discouraging polluters who may be taxed on their excessive emissions, or who must adhere to certain emissions caps to abide by government or treaty protocols.

The carbon in the case of the ocean would be sequestered by phytoplankton. But carbon trading and, in turn, carbon capturing programs haven't caught on as much as expected. No one has been able to figure out a stable, uniform price for carbon, therefore profits have been elusive. Additionally, politics come into play. In the case of the United States, the federal government doesn't incentivize carbon emission reductions.

One entrepreneur, however, wasn't put off by the daunting commercial prospects nor the shade thrown at iron dusting by the marine science community. In 2012, self-professed environmental researcher and eco-entrepreneur Russ George trucked one hundred tons of iron dust to the coast of British Columbia, loaded it up on a fishing boat, and dumped the lot into the Pacific Ocean, about two hundred miles from shore. Environmentalists went ballistic. They accused George of illegal dumping, violating United Nations covenants on geoengineering, and other international protocols. But guess what? What he did worked. The ocean came alive with schools of fish and marine life.

"Our 2012 project was targeted to provide a pasture into which the baby pink salmon of that year were swimming into. Instead of mostly starving, they would be treated to a feast and survive. They would swim home the following year, in the fall of 2013. In Alaska, which catches the vast majority of pink salmon, the prediction, the forecast, was that between 50 and 52 million fish would be caught, and they are never off by more than 5 percent, the forecasters. Instead of between 50 and 52 million fish being caught, they had to stop the fishing, the catch, at 226 million because there was no longer any location on land to accept another fish," George says.

Indeed, the Alaska Department of Fish and Game confirmed a record was set that season. But the same kind of blowback from the marine science community that dogged Martin came George's way even more furiously.

"A week before Easter in 2013 while my team of young people were staring through microscopes, identifying plankton samples and crunching electronic data, a twelve-man Canadian government SWAT team in full body armor with weapons burst into the laboratory in downtown Vancouver and held me at gunpoint on the floor for twelve hours while they ransacked the lab and destroyed our ability to process the data," George claims. He says a vast conspiracy is at work to keep iron dusting from being legitimized. Even the label the marine science community has given it, "iron fertilization," he says is pejorative because it harkens negative associations to synthetic fertilizers, which are, as previously described, extremely harmful to ocean life; fertilizers are what largely create ocean dead zones.

The office raid, he says, "destroyed us." "We had the data that would have proven, would have answered, all of the questions everybody wanted." For example, he says his research discovered exactly how much carbon was removed from the air by iron dusting the ocean.

Okay, so now some caveats and some candor about George. The Haida Nation tribe that worked with him in British Columbia removed him in 2013 from their board overseeing the salmon restoration project. They said in court reports that he misrepresented his credentials and made false and misleading claims about his research. And he does come off in interviews as a bit zany. But the man, who is in his late sixties, provides a sober accounting of his background and how the Haida experiment came to be.

He says he was working as an ecologist at a research institute in Palo Alto, California, in the 1980s when a friend told him about Martin's experiment. It piqued his interest. Later, when he was working in Canada on a carbon sequestration project planting trees, he remembered Martin's scientific finding about phytoplankton and realized it would be far more efficient to revitalize the ocean in order to store carbon rather than to plant trees. That's when he claims the rock star Neil Young, who docked his boat in the same harbor as he, lent him his 101-foot yacht for a test in the Pacific. (There are numerous reports that support this claim, and even a blurry YouTube movie that shows George aboard Young's schooner, *W.N. Ragland*.)

That test in June 2002, George says, was a success. "January of 2003, the first issue of *Nature* for the year comes out, and I'm the feature story in an article called 'The Oresman,'" he says. And he was. It reported his tactics and his call to action, which he repeats: "Listen, it's so easy to save the planet. We just have to restore the ocean to its historic level of health and productivity. The ocean will easily manage the lion's share of CO_2, as it has always in the past. It's dirt cheap, right? You don't need rocket science. You don't need a 50-million-dollar research vote. You can do it on a 125-year-old wooden schooner, under sail. This is an immensely practical method and technology to restore the oceans to health, right? In the bargain you take CO_2 out of the air."

The problem, he claims, is that nobody cares about any of that. Instead, "Everybody cares that 20 percent of one Amazon rainforest has been destroyed." It comes down to, in George's thinking, public disassociation with the ocean.

Iron dusting is a relatively simple process. Iron dust is a common industrial material and can be had at most major hardware supply stores. The size of the particles is what's important: big enough to sink and not get blown away by the wind, and small enough to dissolve relatively quickly. For his experiment, George used fifty-pound bags of red hematite, the oldest known iron oxide mineral on Earth. You can order it from Amazon.

The dust is hauled onto a boat. Once sailed into the dead zone's waters, it is spread just like fertilizer on a lawn. In fact, a lawn spreader device attached to the hull could be used to equally distribute the particles. The dust sinks to the bottom, and then nature takes over: Plankton gobble it up.

"Iron fertilization would be ten to one-hundred times cheaper than

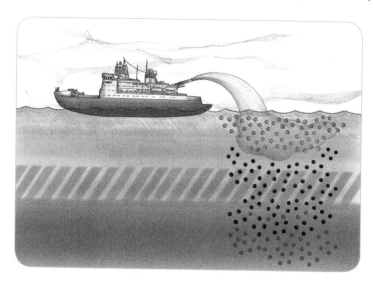

Iron Dusting Oceans

forestation, the next cheapest option," Ulf Riebesell, a marine biologist at the Alfred Wegener Institute for Polar and Marine Research in Bremerhaven, Germany, says in the *Nature* article about George. However, he adds that there isn't enough scientific knowledge about phytoplankton, which "makes it difficult to know which of the possible side effects will occur."

It was this same *Nature* article that found its way into the hands of the Haida Nation, an Indian tribe that calls Haida Gwaii, a group of islands off the coast of British Columbia, its home, and which relies on fishing for much of its diet and economy.

A branch of the tribe who reside in Old Massett Village had reaped more fish than usual after a 2008 volcano eruption in the nearby Aleutian Islands. The eruption had spread dust onto their fishing grounds. In George's experiment, the village saw an opportunity to artificially replicate the volcano's bounty. George was brought on and in short order the experiment was begun. (George had attempted other experiments earlier, in 2007 and 2008, but they were thwarted by environmental activists, including Captain Paul Watson, famous for his whale-hunting interventions.)

For the 2012 experiment, George says all the boxes were checked with regulators and officials. That fact, however, is in dispute, and is what eventually led to the raid on George's laboratory in Vancouver.

George blames the alacrity and aftermath of his "successful" experiment on the UN, commercial actors, and academics reliant on research grants who don't want to see iron dusting solve the carbon emissions problem around the world and eliminate their research funding. As of this writing, George was living in London and was trying to sell his iron dusting solution to other countries.

His big vision is for countries suffering from low fish stocks and dead zones off their coasts to release fleets of ships carrying iron dust that will disperse it in mass quantities. The areas, of course, would be identified, and the dusting well documented. Fish populations and carbon management could then be better measured.

The negatives pointed out by critics of iron dusting, or fertilization, or what have you, include the production of toxins in the oceans, dangerous types of algae, and red tides. The ways in which clouds are formed could also be influenced, and ozone depletion could occur. Oceans are critically linked to all sorts of meteorological phenomena.

With so many in the science and environmental communities against iron dumping, the chances for widespread adoption seem slim. Yet without some type of quick-fix solution, the oceans are destined to keep dying. More dead zones will surely appear on the horizon.

Ocean CPR

It seems like an obvious solution: pump oxygen back into oxygen-starved areas of the ocean. Swedish scientists believe they can do this and offset eutrophication, when a body of water is enriched in nutrients that cause a dense plant growth, which then depletes the oxygen.

Using large pumping stations, the Baltic Deepwater Oxygen project pipes cold, oxygen-rich well water down to the seafloor where it binds with the phosphorus produced from sewage treatment plants, agriculture runoff, and industrial waste. The new, oxygen-rich mixture decreases eutrophication and stops vast cyanobacteria blooms—dead zones.

The project has begun in the Baltic Sea, where a massive dead zone exists.

Swedish scientists began testing ways to bring the Baltic back to life by experimenting on smaller scales in fjords. The experiments worked and the Swedish government is backing the ambitious Baltic project, which expects to fully restore the sea within fifteen years.

The Baltic Sea is bounded by Sweden, Finland, Russia, Estonia, Latvia, Lithuania, Poland, Germany, and Denmark. It is notoriously brackish because it is mostly enclosed and pollutants can't easily wash out into larger bodies of water, such as the Atlantic Ocean, for dispersion.

The Baltic Deepwater Oxygen project, or

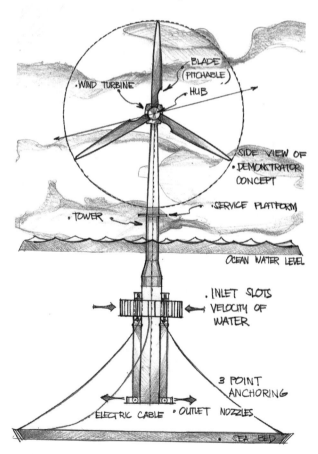

The Baltic Deepwater Oxygen Project

BOX, as it is referred to, uses a floating wind turbine unit equipped with pumps for oxygenation of the deep water. It is the best hope for ocean-cleansing dead zones, scientists believe.

Other methods have also been suggested, such as using chemicals that bind the phosphorus to bottom sediment to stop growth. Or dredging the sediment itself in order to disperse phosphorus. But oxygenation seems to be the big winner.

Still, BOX has its critics. There is concern that geoengineering the ocean depths could interfere with marine life struggling to adapt, and damage them further. The new water could raise deepwater temperatures and do the opposite of what's intended and expand dead zones. But the biggest salvo is that BOX only addresses the symptoms of eutrophication, not its causes. Polluting less is the ultimate solution. And it has been tried. Still, the algal blooms grow and grow. BOX may be a quick fix, which could be better than no fix at all. If polluting the Baltic goes undeterred, eventually the majority of the sea could be left virtually lifeless.

The BOX project involves one hundred floating platforms strategically placed over or near dead zones. It may change how the surface of the sea looks. But that may be a small price to pay for the damage underwater that has already been done.

Dead Zones As Fuel

Researchers at Stanford University in California have figured out a way to clean up ocean dead zones and at the same time turn the waste into energy. It seems like a smart idea, producing twice the benefit. It was hatched by rocket scientists who knew that nitrogen, a major hazard for marine life, was also a major component of rocket fuel.

Here's how it works: The wastewater from the dead zone is skimmed and put into a tank. There bacteria break down ammonia from the wastewater. Ammonia contains nitrogen. The ammonia is transferred to a reactor that injects oxygen into the mix and produces nitrous oxide. Nitrous oxide (often referred to as laughing gas) is a greenhouse gas, but it can also be burned as a fuel—rocket fuel.

The rocket fuel produced can power the entire cleaning system while also extracting a huge, new source of renewable energy. In this way, it's self-generating.

A successful test pilot of the cleaning system at a northern California wastewater treatment plant proved the process works outside a laboratory environment. However, scaling up to meaningful size is the challenge.

There are approximately four hundred areas suffering from eutrophication around the world, comprising an estimated one hundred thousand square miles. That amount of seawater would produce a heck of a lot of rocket fuel. But to clean it all is an impossible task, which is why sights are being set on wastewater treatment facilities on land. Almost every sewer system has a wastewater treatment plant at its end that removes harmful contaminants before water is released back into the environment. Pollutants collected at coastal plants could be converted into fuel instead of being washed out to sea. That in and of itself could also help bring oceans back to life. The nitrogen from wastewater treatment plants (along with other toxins) is what helps to cause dead zones in the first place.

CHAPTER 9

Dutch Sea Level Defense

The most vulnerable place on Earth for floods isn't easy to pinpoint. Floods are the most common natural disaster, doing more damage to the planet every year than any other type of weather event, including blizzards, tornadoes, and typhoons. They produce billions of dollars in damages across the globe, affect millions of people, and can spring up nearly anywhere, even in deserts. If flood risks, which are on the rise due to climate change, aren't lessened, it's predicted that floods will inflict more than one trillion dollars of damage annually on the global economy and put at risk major population centers.

By some counts, Guangzhou, China, a major commercial hub two hours northwest of Hong Kong, has the highest risk of floods because flooding there can cause the highest amount of economic damage and make vulnerable many of the city's nearly fifteen million residents. By other measures, the combination of river and coastal floods in India make places such as Calcutta and cities along the Bay of Bengal—squeezed between the ocean and the high mountains of the Himalayas—most at risk. The World Atlas lists the tiny country of Suriname in South America as the area in which the population could be almost entirely consumed by floods because it is so low-lying and has weak infrastructure. But ask the government of the Maldives, which is composed of more than one thousand islands no more than six feet above sea level, and they will say that they are most exposed. As testimony to their anxiety about being washed away into the Indian Ocean, the government held an underwater cabinet meeting to highlight its fears.

And yet, by measure of percent of its population living in flood zones and by sheer historical association, the Netherlands, whose name literally means "the low countries," stands apart as the country that has lived with washout issues for centuries. Half the people in the Netherlands live in critical danger of flooding. Holland, the more colloquial name for the Netherlands, is as much as twenty-three feet below sea level—matching Denmark with the lowest land points in western Europe—and it suffers from widespread inclination, or sinking. That should be no wonder. Many areas of Holland have been reclaimed from the sea. Without dikes and levees, those reclaimed areas would be indistinguishable bodies of water. Such is the country's relationship with land and ocean, ebbing and flowing. It holds its place like a fierce old mother, to steal a few words from the poet Walt Whitman, and gives structure to that which is not sound.

By magnificent acts of engineering, the Dutch have managed to hold the sea back for centuries. The oldest dike in the Netherlands dates back two thousand years. It was built by monks using turf. The first dikes were made to protect cropland, and then people started making bigger mounds on which to live. In the north of Holland, villages eventually connected their dikes. This created an embankment, the Westfriese Omringdijk, seventy-five miles long. It's a remarkable structure that stands strong even today.

Around the fifteenth century, windmills became popular in Holland and were used to pump water out of sunken areas. (Hence the famous association between Holland and windmills.) However, it wasn't until the late eighteenth century—when the Dutch Ministry of Infrastructure and the Environment was created—that construction and maintenance of waterways and water management began in earnest.

Twentieth-century technology brought about another new beginning for the Netherlands when in 1918 a massive public works program was begun to barrier the North Sea to protect agricultural land and improve water management through a series of hydraulic engineering projects. By dredging waterways, damming inlets, and draining lands, engineers architected what some consider one of the seven wonders of the modern world.

But there are new concerns, twenty-first-century concerns, that the system of locks, dikes, and levees will fail. Dutch engineers typically design for two hundred years into the future. Extreme weather and sea level rise have distorted past plans. The risk of damage associated with floods over the coming decades has now doubled. That has Dutch water officials anxious, if not panicked.

As global temperatures rise, more water is allowed to loft about in the air (warm air holds more humidity than cold). When that extra water is shaken loose by storms, more precipitation falls onto land than can be absorbed. Flooding ensues. In geography such as the Netherlands, where land is already water-laden, this means floods happen more easily. Sea level rise causes coastal erosion, as well. With rising ocean temperatures expanding ocean mass and glacial melt adding to volumes, a higher sea level is inevitable. By the year 2050, sea surfaces could rise between two inches and a foot—and that is by the most conservative estimates. High estimates warn of an eight-foot rise by 2100. For a nation that largely exists below sea level, that could spell ocean eclipse. Every inch of sea level rise equates to ten feet of beach loss, and flooding chances rise exponentially, to as much as one hundred times past occurrences.

Extreme flooding events are already unfolding. All over the world, more one-hundred-, five-hundred-, and thousand-year floods are occurring every single year, canceling out the way geologists have figured and projected flood events. So much so that the Dutch government has committed 1.5 billion dollars annually through the year 2032 to fortify its water defenses.

It may not be enough.

Standing on the beach in Almere, twenty-one miles southwest of Amsterdam, and looking out at the bay, then looking up at the cloudy sky, perhaps brings about the same feeling as what Hans Brinker's little Dutch boy, who you might recall plugs a dike with his finger, felt: stuck in a bad spot. Almere is the newest city in Holland. It was reclaimed from the bay fifty years ago and a planned community of more than two hundred thousand people now live in the area. The topography is strange. Small canals cut through housing complexes and switch back underneath roads. Little bridges abound. Swamp grass pushes right up to the sides of extremely modern-looking homes and tall, eccentric commercial buildings. Plazas unfurl to quaint waterways and ponds. Land cuts this way and that and carves in sharp angles. It doesn't feel right. It doesn't feel solid.

Even without the portents of climate change, Holland is in a tenuous place. Add a climate push and calamity becomes all too real. You don't have to look far back in time to find examples of how the country might suffer.

In 2017, as much as one-third of Bangladesh was flooded in a single rainy season. Forty-one million people were affected; homes, commercial structures, and agricultural land was washed away. Twelve hundred people died. At about the same time, Hurricane Harvey smacked into the southern United States, producing record rains and massive floods in Texas and Louisiana. It caused 125 billion dollars in damages, and flooded three hundred thousand structures and half a million vehicles. Eighty-two people were killed. In 2018, record rains and floods pummeled Argentina and Chile. And in the same year, Somalia was inundated with never-before-seen precipitation, forcing 175,000 people from their homes.

By the time these words are read, there will likely have been even more catastrophes to point to. These are not isolated phenomena. They are part of a pattern. And that is what has the Dutch so on edge. A significant flood disaster on their shores could wipe away much of the country and beyond.

A January 2018 study of flood risks by European scientists set off alarm bells throughout the continent. It found that global warming is linked to a substantial increase in flood risk over most countries in Central and Western Europe at all warming levels, even the most optimistic scenarios, where temperatures rise just 2.7°F from preindustrial levels. Temperatures could more than double that estimate.

Over the twenty-year period from 1995 to 2015, about a third of the world's population—2.3 billon people—have been impacted by floods, and the costs have mounted into the trillions.

It's hard to imagine things getting worse. But engineers and architects are paid to imagine worst-case scenarios. That is why floods are often defined by a statistic. We regularly hear of those one-hundred-year, five-hundred-year, or thousand-year floods. The numbers don't mean that such a flood will occur once in a hundred, five hundred, or thousand years. It means that there is a one-in-a-hundred, -five-hundred, or -thousand chance of such a flood happening during the course of a year. It's all probabilities. And Dutch engineers and architects don't like the odds they are facing.

A series of events dubbed the "worst floods in decades" serially hit Europe in the years leading up to events called "the worst floods in a century" in 2016. "Worst" seemingly takes on new meaning every year. No one wants to wait around and get caught unprepared for the big one—a paleo flood of biblical proportions.

In the Bible, Noah's Ark was built to withstand such a great flood. The flood changed the landscape of the world and forced—if you believe in the Good Book's narrative—species of all kinds to board the ark in order to escape destruction.

Scientifically speaking, great floods did occur with regularity in the long-ago past, in Paleolithic times, hundreds of thousands of years ago.

The Paleolithic period lasted from 2 million B.C. to 10,000 B.C. Paleo floods, scientists found, came after long periods of drought. Atmospheric rivers, which are invisible bands in the sky that transport water vapor around the world, opened up and deluged entire regions. These superfloods occurred about every two hundred years, according to climatologists, and their pattern, these scientists believe, could begin again.

In late 1861, a megaflood turned part of California into a three-hundred-mile-long sea. It rained hard for nearly two months, covering Sacramento, the state's capital, in ten feet of water. A twenty-mile-wide lake was formed in the Central Valley, previously dry land. Mudslides ruined communities. One in eight houses was destroyed. The flood reconfigured much of the state. Do the two-hundred-year math; another megaflood could be due.

The United States Geological Survey (USGS), a government agency whose job it is to track floods, droughts, and the like, has devised a name for the so-called inevitable megaflood: ARkStorm. It's obviously a take on the Noah's Ark story in the Bible. But it is also a sly acronym. The "AR" stands for atmospheric river. The "k" is for the one-in-a-thousand-year event. And "storm" is, well, obvious.

The ARkStorm scenario is referred to as California's Other Big One. The Big One is the anticipated earthquake that will arise from the San Andreas Fault and shake California like never before, perhaps even crack it loose from the continental shelf. "A severe California winter storm could realistically flood thousands of square miles of urban and agricultural land, result in thousands of landslides, disrupt lifelines throughout the state for days or weeks, and cost on the order of $725 billion. This figure is more than three times that estimated for the ShakeOut scenario earthquake, that has roughly the same annual occurrence probability as an ARkStorm-like event," the USGS claims on a web page it has dedicated to the epic-flood hazard scenario. It estimated 1.5 million people will need to be evacuated. The ShakeOut scenario is the destruction linked to the Big One.

Clearly, the ARkStorm would be devastating for the West Coast of the United States. But what about extreme flooding potential in other areas around the world?

Different reports have New York City subject to extreme flooding every five years in the very near future, the number of Paris floods doubling, as well as swaths of Asia, Africa, South America, and central Europe completely drowned.

Floods come in different types, and nearly all are predicted to be more severe in magnitude on account of climate change. River floods are usually caused by excessive rains, snowmelt, or ice jams that back up running water and cause it to overflow. A coastal flood is caused by higher-than-average tides and winds that blow water onshore. Storm surge, produced by low-pressure systems, can also cause coasts to flood. Inland flooding happens with intense rains—when land cannot absorb water fast enough. Flash floods can happen in minutes whereas other floods can gestate over days. No place is really safe from floods, except for perhaps the McMurdo Dry Valleys in Antarctica that reportedly haven't seen rain for millions of years.

The lowest point in Holland seems like no more than a dip in the road. That's because that is exactly what it is—a twenty-three-foot depression adjacent to the A20 highway just outside of Rotterdam. It's only when you drive back up to flat surface that you realize you were below sea level; canal waters match the roadway. The threat of flooding and the scare of sea level rise are faced dead-on.

The Netherlands is a series of waterways, islands, bridges, and beachheads. Nearly four thousand miles of winding canals make their way through the small nation, which isn't even twice the size of the state of New Jersey. Eighty-four locks and 278 bridges keep things moving—freight and people. Cycling is the major mode of transportation, but water is everywhere, hardly ever out of sight.

Zuidplaspolder is the name of Holland's lowest land point. Nothing about it belies the fact that flooding and sea level rise are threats. Water seeps onto roads. Marshes and wetlands surround it before giving way, at least on one side, to the North Sea. A giant ruler marks the spot, and a quaint drawbridge jets over its canal, designed as a ring.

Bridges in Holland are something to behold. They come in all sizes, and many are stellar works of towering art, suspended in different types of designs. Arcs, serpentines, and elbow bends describe just a few of their forms. The Erasmus Bridge in Rotterdam is of particular note. It is shaped like a swan and rises nearly five hundred feet over the Maas River, connecting the north and south parts of the city. Long cables stretch out to a pylon. It is more than twenty-six hundred feet long and is movable.

Bridges play a critical role in the Netherlands. Half of Holland is less than one meter (about three feet) above sea level, necessitating bridges to get from one point to another. Land's relationship to sea level, namely at coasts, is important because it signifies how vulnerable an area is to erosion, flooding, and being consumed by the sea. Most centers of commerce and population are coastal, closer to centers of trade. This plays out worldwide. Forty percent of the global population lives in coastal areas. Sea level rise could swamp these areas, resulting in massive population displacement, infrastructure loss, economic loss, water contamination, and other catastrophic possibilities. This is why flooding, already a natural concern, has become such a worry.

As a boy growing up in rural Holland, Kristian Koreman was keenly aware of floods and the environment. He had to cross a canal to get to school every day. If the canals swamped out, he'd miss class. The circumstance shaped his future. "Ever since I was a boy, I knew I wanted to be a landscape architect," he says, gazing at a completely artificial community he designed abutting Almere.

Koreman and his partner, Elma van Boxel, founded ZUS, or Zones

Urbaines Sensibles, named after what it creates: sensible urban zones. The landscape architectural firm has become a force in the design community because it has a very distinct philosophy of working with a location's assets rather than against them. "Re-public" is what ZUS defines as the architectural mandate of the future. It's a belief that space and politics need to feed off each other to engender better quality-of-life issues (how to use and plan public spaces, for example), as well as to create welcoming and inclusive environments (disallowing segregation, as another example). It is this thesis that is the reason why ZUS is highly sought after to plan for environmental as well as social consequences. For instance, ZUS is helping to turn New Jersey's Meadowlands into a new place of opportunity and redevelopment. Of course flood protection is foremost, but ZUS is also taking into account residents, visitors, transportation needs, and recreational offerings.

Across the harbor, Koreman and van Boxel run a gentrification lab in Brooklyn to facilitate resident integration and are working on other projects in South America. Commissions are worldwide. However, it all began when van Boxel and Koreman met in architecture school in Holland and decided to focus on creating a new urbanism. One of their first designs was in 2005 for a floating city called the Tide City.

"How are we to survive in a delta when the sea level is extremely high? The Tide City study suggests that instead of withdrawing fearfully behind the dikes, people should live in and with the delta. Inspired by the powerful dynamics of the current and the tides, the town would float on the water like a piece of flexible seaweed. It would be composed of large, medium, and small pontoons linked like tentacles to a central square. A [four-hundred-foot] long bridge would connect the square to the mainland and ensure that the structure can move up and down. Tide City would generate a very location-specific experience with a constantly changing view. Above all it would provide a safe place to live, with opportunities for land reclamation and energy generation." That is how van Boxel and Koreman describe the project that to date exists only in model stage. But it turns out it serves another purpose. The Tide City laid the groundwork for the Delta 3000 project.

Delta 3000's scheme ambitiously calls for low-lying areas throughout the country to be covered in sand, and man-made hills to be added to the landscape. Covering the country in sand would prevent flooding, produce fresh water, and create a naturally sustaining ecosystem. It might even turn the Netherlands into more of a Mediterranean beach resort location and create more space for future climate refugees who will be forced from their lands.

"We wanted to reestablish the connection of living in a kind of natural landscape. Through technology we can reestablish that natural system again," Koreman says, taking rare time out of his day to show a visitor around the spot of artificial dunes next to Almere that he designed. Three thousand homes are being built and companies such as

The Delta Project

Carnival Corporation are moving in. Thirty years old and hip in the fashion of a Berlin or Brooklyn creative (black wardrobe, sneakers, and contemporary, intellectual glasses), Koreman allows for a bit of free-spirited sprinting up and down the sandy dunes, breezily accommodating a request to visit the depths of the construction site.

By dredging Lake Almere and eventually the nearby North Sea, there would be enough sand withdrawn to heap billions of tons onto polders, or reclaimed land areas. These dunes fare better than traditional groundwater pumping and historical seawall battening, which are common throughout Holland, because the dunes actually strengthen soil rather than weaken it. When groundwater is siphoned, it lowers water tables and causes soil to sink. Moreover, sandy dunes purify rainwater naturally and can keep saltier groundwater at bay. ZUS's project makes foundations firmer with the added benefit of aesthetic appeal.

The sandstone homes that dot the dune community on Almere's waterfront are handsome. The mounds afford privacy without obstruction. Built in rows that twist in sections of a half dozen or so townhouses, the plots blend nicely together. It is reminiscent of Cape Cod except for the architecture, which reflects more of a Mediterranean style. A model unit reveals wood floors, stainless-steel appliances, whitewashed rooms, and sliding glass doors that open to a small, naturally landscaped yard. Most of the houses are two levels. Koreman says selling the homes is part of a bigger strategy. "If there's housing and real estate to be developed, then it's possible to build a business case where those private investments lead to this big building investment of a dune," he says.

The Delta 3000 project, named as such because it is meant to secure the Netherlands for the next millennium, to the year 3000, will cost billions of dollars and will need private sector participation to fund it.

Mechanically, the Delta project involves dredging, hauling, dumping, and compacting tons of earth. Construction is complex. The new ground shifts before it settles. Foundations, plumbing, and utility service piping have to account for all that movement. Utility infrastructure is typically not very forgiving, and a look at the manholes that appear every few feet shows just how unsettled the ground is. Workers need quick access to fix and replace water and sewer lines. "It's just part of dealing with reclaimed land," Koreman explains.

Certainly, Almere is not alone in being raised from the depths of the sea. Twenty-five percent of Singapore was built from reclaimed land. Mumbai would not exist as it does today if engineers hadn't figured a way over the course of five hundred years to link together the tiny seven islands of Bombay. Major sections of the San Francisco Bay were reclaimed and built upon. (So much so that underground streetcars reportedly pass through the hull of an old sunken ship buried underneath the city.) Tokyo, Rio de Janeiro, and the capital of New Zealand, Wellington, are further examples of either total or partial land reclamation. Taking it a step further is the United Arab Emirates, which is constructing entire fake floating cities. They are being designed as "best of" replicas: three hundred artificial islands that are made to look like various places around the globe, including Monaco, Germany, Sweden, and cities such as Venice, Italy, and St. Petersburg, Russia.

The dune city of Almere has no such replica resort-style ambitions. A practical design issue is being forged. Dunes are land's first climb out of the water and can be modeled to save society from the ruins it has made. Or at least Koreman sees it that way. And while it's heartening to see and hear such calls for natural and societal symmetry, even the best intentions have their downsides.

There is no getting around the fact that years of mega-dredging the seafloor will have to occur in order to build new lands and new cities and communities in Holland for the Delta 3000 project. Dredging, however, brings about numerous environmental concerns. When seafloors are disturbed, marine life can be harmed. Materials previously sunken and deposited get swept up into the sediment and spilled about. Sediment spill can prove toxic, and heavy metals from it can be released into waters, where the toxins can in turn be consumed by fish. Mercury contamination is already a major problem for marine animals. And mercury poisoning can cause brain defects, heart disease, and strokes in humans. Other heavy metals such as copper, zinc, and cadmium also bring about ill effects in sea and on land.

Koreman says environmental testing shows that no harm comes from dredging for the Delta 3000 project. Still, it will take millions of tons of sand to complete the 320-acre project in Almere alone, and another twenty-two billion tons for all of Holland. That's enough in dump trucks to circumnavigate the planet a thousand times.

Digging and dredging and dumping is sure to have some type of environmental effect somewhere. Is it worth disturbing nature so? Or is it wiser to cue the signals the elements are showing and allow the seas to lay claim to what is increasingly theirs—dry land?

Dr. Jimmy Jiao, an expert in land reclamation matters and a professor of earth sciences at the University of Hong Kong, believes land reclamation should be avoided as much as possible in most places. There are exceptions, he explains, in cities like Hong Kong, where land is really limited. But even in places like that, comprehensive testing and analysis may prove that certain, very specific areas are better candidates for filling than others. Groundwater mixing that affects drinking water, for example, is one major concern. The possibilities of releasing chemical compounds housed beneath seabeds is another. In addition, the types of fill materials used can carry all sorts of effects. He points to a raft of research he has produced on the subject of land reclamation and the extreme interconnectivity of elements involved with coastal ecosystems. As in the game of chess, or maybe more appropriately, the game of Go, moving one piece on the board can set off a mind-numbing relay of sequences. Disturbing sediments on seafloors can produce innumerable consequences. We are, it seems, damned if we geoengineer in some cases, and damned if we don't in others.

If the high estimate of sea level erosion comes to fruition, thousands of miles of coastline around the world will be lost. Flooding could prove to concentrate populations in areas farther inland and outside flood zones. Moving such huge portions of civilization won't be easy.

According to a highly academic paper by Nobuo Mimura, the director of the Institute for Global Change Adaptation Science at Ibaraki University in Japan, "The sea-level rise projected for 2100 poses significant threats to coastal zones in the world. Particularly, when the intensification of tropical cyclones is superposed on sea-level rise, the population at risk from inundation is likely to amount to several hundred million [people]."

The idea that we are faced with the very real possibility that sea level rise will affect a population about the same size as lives in the United States is difficult to believe. Flooding in the Netherlands is one thing. That is easy enough to comprehend. But a massive displacement and decimation of people and property is something that has been perceived as merely an outlier of academic theories—something to contemplate but not really consider. Until now. Now it's all too real.

To avoid massive flows of climate refugees, we are going to have to produce more land on which people can live. Artificial cities, born from the depths of the seas, could be where many of us will call home.

Floating Cities

Waterstudio.NL is designing for living in the future—on water. "The prognoses is that by 2050 approximately 70% of the world's population will live in urbanized areas. Given the fact that about 90% of the world's largest cities are situated on the waterfront, we have arrived to a situation where we are forced to rethink the way we live with water in the built environment." That is the vision of the firm led by Koen Olthuis, who is considered a leader in the design world. His projects range from entire floating cities to floating homes, floating golf courses, floating restaurants, floating cruise terminals, floating mosques, floating hotels, and even floating slums. The projects are stunning feats of architecture all aimed at the common mission of planning for the inevitable flooding consequences that climate change will bring.

Waterstudio.NL is based in the Netherlands and describes itself as "architecture, urban planning and research in, on and next to water." Designs are not only for humans.

The Sea Tree is designed as a vertical green habitat for animals only. It is radical looking: open air "floors" that stack, becoming wider and longer as they go higher. Vegetation spills out from each, as the model for it shows. The entire structure floats on water, and it is meant for urban environments as high-density green areas within cities are sacrificed for development.

The Sea Tree is built on the same technology that is used for offshore oil rigs, and the hope is that oil companies themselves will donate a Sea Tree to a city of their choosing to express their concern for the environment. The Sea Tree is the first floating tower built exclusively for flora and fauna.

For us humans, Waterstudio.NL is developing architecture for virtually all aspects of our lives. The Westland is a floating city being designed next to The Hague. It incorporates social housing, floating islands, and floating apartment buildings.

Mindful that the poorest among us are on the front lines of climate change and are most susceptible to its destruction, Waterstudio.NL has designed Wetslums, floating sea-freight containers built on top of foundations made from PET bottles. The units are super modern-looking and stark-white inside. A network of UNESCO hydraulic engineers is being tapped to help install the developments worldwide. The project is part of Waterstudio's Floating City Apps Foundation, whose goal is to upgrade the living conditions for waterfront slums.

Olthuis believes today's designers are an essential part of the climate-change generation and should consider more dynamic solutions to ur-banization, solutions—by his vision anyway—that will take into account a world increasingly shaped by water.

City in the Sky

"The concept is inspired by the Lotus flower which is known for its ability to emerge above the murky waters pure and clean." That is how Tsvetan Toshkov, the Bulgarian-born archi-tect and visual designer who now works and lives in London, defines his

Lotus Flower City Towers

city-in-the-sky renderings. Toshkov is the founder of studios that combine architecture, computer graphics, and design in ways that are visually stunning.

The beautiful lotus flower glass structures stretch high above skylines and soar through clouds. They beckon natural splendor. They promote environmental thinking and evoke a quiet sense of respite in an otherwise chaotic urban abyss. Their utility may be far more than aesthetic in the future; they could add much-needed space on Earth for the expected population explosion this century.

Bucking a dystopian blueprint, the rising lotus towers are transparent glass, allowing for sunlight to be captured and views to be unobstructed. The lotus leaves at the top spread out. These green spaces—covered gardens and ponds—are featured in a fashion meant to breed tranquility. Living, working, shopping, entertainment, and educational spaces are all built into the layers below.

The design was inspired by the lack of green space in downtown Manhattan. The lotus towers are designed with nature and sustainability in mind. They are meant to be constructed in megacities, populations of more than ten million people. They could be built in bunches and would have leaves that touch, providing connections from one "city in the sky" to another. Of course, their height would also mitigate flood exposure for all but the bottom floors.

Megacities are increasingly building higher structures. The world's tallest building, at more than 3,280 feet, will be the Jeddah Tower in Saudi Arabia when it is completed. It will top by more than 500 feet, the Burj Khalifa in Dubai.

Cities going vertical is nothing new. But city-like structures are. A nonprofit organization, Vertical City, has even been launched to "ignite a worldwide conversation about vertical cities as a solution to a more sustainable future." Some cities have a head start. Hong Kong has more than 350 skyscrapers; New York City has more than 275; and Dubai's skyscrapers eclipse 190, with dozens more under construction. These may not look like lotus flowers. But Toshkov's city-in-the-sky concept is blossoming.

Floods may prove it more than an architect's aesthetic indulgence. Vertical cities may be necessary for urban life and for commercial and residential developments to grow—up.

CHAPTER 10

Living Beneath the Surface

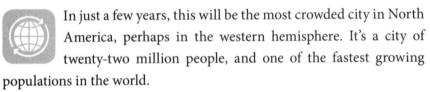 In just a few years, this will be the most crowded city in North America, perhaps in the western hemisphere. It's a city of twenty-two million people, and one of the fastest growing populations in the world.

Mexico City is adding more people than it has space for them to live. There are few tall buildings. Zoning restrictions abound. And it is quickly running out of land on which to expand. Geographic boundaries block it in. This is a metropolis bordered by mountains and volcanoes— the Sierra Nevadas to the east; and lower sierras and ranges to the north, west, and south.

For millennia, hordes of people have been drawn to this area because of its wealth of natural resources—pine forests and rivers, wildlife, and even saltwater lakes. There is food and water aplenty. But now there is no more room to grow.

Tokyo, with thirty-eight million residents, is the most populated metro area in the world. It has grown upward, with skyscrapers, and outward without the impediments of hard natural boundaries. Similarly, New York City has jumped the East River and morphed into Brooklyn and Queens and beyond. London has crossed the Thames and has headed through South Bank. For Mexico City, there is no such luck. Nature traps people in its valley. And Mexico City is on track to add millions more people by mid-century. Goldman Sachs says it may become the world's fifth-largest economy by then. Growth prospects of all sorts are predicted. Still, historic preservationists ban renovations. Zoning laws won't allow developers to build anything taller than five stories in the city's core. Transportation infrastructure is aged, and the surrounding mountains inhibit connections to satellite towns that could serve as commuter alternatives. The result is mass urban concentration.

Walk through any megacity these days and the crowds are sometimes difficult to fathom. You get swept along in them. The only place to step is the space left behind by the person in front of you. Shoulder to shoulder, there is no way to break free.

Getting caught in a crowd is something most of us have likely experienced. It can be overwhelming, a freak-out moment, for sure. Which is why standing alone in the middle of Mexico City, in the center of its main square—Plaza de la Constitución, or the Zócalo, as it is commonly called—is so surreal. The plaza stretches nearly eight hundred feet in each direction, and is cordoned by national monuments. Yet, there are only a few dozen people at a time crossing it. It is a vast, open area, just a lone flagpole rising high in its center.

Mere steps away, outside of this main square, the city is packed, jam-packed with people and vehicles and public transportation congestion. There are organ grinders and tour groups and painted-face performers in near-naked native garb. There are street stalls that extend for miles and miles. Hordes of workers and students and more police than you'll probably ever see anywhere else move about the city

With as many as ten million more people poised to be added this century to the already overstressed urban matrix, and with space limitations, the question is, where will everyone go? In all likelihood, down, underground. And they won't be alone. All around the world, people live underground, entire cities even.

In Coober Pedy, Australia, it's so hot that people who moved there a hundred years ago to prospect for opal realized they had to build their city beneath the surface of the Earth, where it's markedly cooler. Now close to two thousand people live there.

Coober Pedy's average annual temperature eclipses 85°F, and during summer remains above 90°F and often more than 100°F for weeks on end. Underground, at depths of more than thirty feet, temperatures remain relatively constant, and cooler. Fun fact: The average local year-round surface temperature is what you will find at these depths, no matter where you are on Earth. In Coober Pedy, that can mean a swing of 20 degrees—a big cool-off.

There are, of course, exceptions to finding reprieve below the surface. The place that holds the record for the highest average annual temperatures is Dallol, Ethiopia. The constant 94°F surface air temperature there won't be tempered by digging down; under the surface lies an active volcano.

In primitive times, caves did the trick for providing cooler environments and protection from the elements. In modern clusters of civilization, living underground seems like an extraordinary step. That is until you consider this: Global temperatures could spike so much over the next one hundred years that it will be nearly impossible for billions of people to remain living aboveground; half of the land area of the planet may become uninhabitable.

A hothouse scenario such as this would mean temperatures rising about threefold more than expected—a lot, for sure, but not outside the limit of possibility. One scientific study states that humans cannot survive more than a few hours if our skin temperature exceeds 95°F in wet bulb terms. Wet bulb temperature accounts for humidity. It is coined as such because it is the temperature of a lightbulb covered in a wet cloth and exposed to air. Wet skin, caused by perspiration, is meant to cool our bodies. Above 95°F, that doesn't happen effectively, and hyperthermia sets in.

According to a study by researchers at Purdue University and the University of New South Wales, Australia, regions too hot to survive if temperatures rise three times more would include most of the East Coast of the United States, all of India, most areas in Australia, and heavily populated parts of China.

With heat and population concentration in mind (people in affected areas would in all likelihood move to cooler geography, congesting those regions), the simplest solution may be underground living. Besides Coober Pedy, there are major subterranean developments already in the works.

In Fukuoka, Japan, architects from the Taisei Corporation of Tokyo have plans for Alice Cities—airy underground spaces connected by subway trains and subterranean roads. In Helsinki, Finland, there is a "shadow city" with a public swimming pool, shopping areas, a church, hockey arena, and an industrial center. In China, old war bunkers have been taken over by people in Beijing. In Singapore, there are plans for an underground "science city" where more than four thousand people will live. Toronto already has its PATH system: pedestrian walkways that span for nearly twenty miles and connect transportation hubs to restaurants, shops, and other commercial spots. New York is even considering a Lowline, akin to its popular High Line park, where pedestrians meander along an old rail line and experience greenery above the mayhem of the city streets. The Lowline is the opposite: It is billed as the world's first underground park. It will use new solar technology to transform a trolley terminal into green space. Breakthroughs in lighting, mood, and spatial aesthetics provide the world with these new underground layers of living and working possibilities.

Visionary architects are teaming with urban planners to come up with living solutions to looming climate change conditions. They've partnered before.

In the past, far-out scenarios prompted considerations for building an alternative world. During the late 1960s and into the 1970s, the Italian architectural collective Superstudio crafted designs for a world scant of resources. The group's architects made blueprints for movable cities—a Continuous Production Conveyor Belt City—among other plans deemed problem-solving. These would be entire communities, or "cities," that live aboard stations that move about to consume as many natural resources as possible. In 2016, Rome's MAXXI, the National Museum of 21st Century Arts, exhibited Superstudio's work, including its 12 Ideal Cities and other concepts that are tragically relevant today and will likely be even more so in the future. The exhibit showed drawings, photographs, videos, and design objects—all of which are fundamentally a critique of society, whether alerting to dangers of overconsumption or warning of the fallibilities of modern infrastructure.

Superstudio architects sought plans for a new world existence. Designs were meant as parables for a world they saw as increasingly headed for the brink of destruction. Today, architects are practically rather than theoretically designing for a world where pollution and temperatures constrict movement aboveground. Additionally, they view urban sprawl as the enemy. Sprawl destroys natural resources and natural habitats. Horizontal planning, where cities broaden their footprint and consume more plots of land, diminishes the amount of local resources available; forests, for instance, are sacrificed for housing and commercial development.

As far back as Leonardo da Vinci's era, big thinkers have made plans for overcrowded living circumstances and ways to build or funnel natural resources to the urban masses. (One of da Vinci's marvelous plans was to divert rivers.) But da Vinci, who died in 1519, and urban planners of those times, didn't have to worry about land space so much. The population explosion hadn't begun. There was plenty of land on which to spread out.

It wasn't until the nineteenth century that the population explosion began, doubling the world's population to two billion people over the course of one hundred years. It has obviously increased exponentially since; there are nearly eight billion people in the world today. Along with more people goes more land. By 2030, urbanized land will comprise three times as much space on the planet as it did in 2000, according to the National Academy of Sciences. China is expected to see its urban areas grow most. And the mid-latitudinal region of Africa is also expected to see bigger cities, followed by urban areas in South America.

The most populated cities in the world as of this writing were Tokyo, as previously mentioned; Delhi; Shanghai; São Paulo; and Mexico City. By the end of this century, the most crowded city in the world will have nearly three times as many people living in its metropolitan boundaries as present-day Tokyo. Lagos, Nigeria, will house 88 million people (it had 20 million residents as of this writing). Kinshasa, Democratic Republic of Congo, will be the second most populated city on Earth, with 83 million—a massive increase from the 11 million people who live there today. Dar es Salaam, Tanzania, will see an even more dramatic rise: 74 million people, up from 4.5 million residents. Mumbai will have 67 million people, three times as many as who live there today; and Delhi's population is expected to more than double to 57 million.

Because these cities are already overcrowded, there is scant room for more inhabitants. Which means localities will be far more congested and resources will be stripped or overstressed.

To combat overcrowding and urban land expansion, a so-called smart cities movement has begun. A smart city uses technology to better manage its resources. Microsoft is a leader in the smart mission with its CityNext program—innovations that map out a more digitally oriented city life by connecting disparate bits of information to do all sorts of things, from managing traffic to transporting food and water efficiently. Google is also a smart city developer through its Sidewalk Labs division. Climate change, no doubt, will affect more aspects of urban life, and managing its effects will be big business in the future.

In past times, without technology such as artificial cooling or heating to solve people's climate adaptation problems, resiliency was had by re-engineering natural structures as human habitats.

Petra, the ancient city in Jordan made famous by *Indiana Jones and the Last Crusade*, was once a bustling center of commerce. Its petrified (hence the name) structures were carved into mountains, and an estimated twenty thousand people lived there in the fifth century B.C. Major underground cities also existed in China, Turkey, Poland, Italy, and Africa. The reason for living underground then, in addition to a more temperate climate, also had to do with defending against raiders and wartime enemies. The climate is actually now considered one of our biggest enemies. The US Department of Defense even lists it as a top threat.

With global temperature rise and absent a silver bullet that cools the atmosphere, it looks like we will have to go underground for at least part of our days, as we have done before. It is a return to the cave, a womb of safety from the outdoors. But these sites need not be crypts.

In ancient Mayan culture, cities were built on top of one another. The Aztecs built their temples on top of a lake, and then after the Spanish conquest, the Spanish built their temples on top of the temples of the Aztecs. And eventually the whole Spanish colonial city was built on top of the Aztec city. That city is now called Mexico City.

The Zócalo here is either a massive open area meant for congregating and celebrations, or it's a giant waste of space, depending on how you look at it. Mexican architect Esteban Suárez looked at it differently. Why not, he thought, take a cue from the past, the once–Aztec capital Tenochtitlán, and build on top of the cities that form the foundation for Mexico City itself? Why not build down, not up? He designed the Earthscraper, the inverse of a skyscraper, that would plunge nearly a thousand feet below the surface, jettisoning past ruins and relics, and reaching back in time, metaphorically speaking anyway.

"We thought it would be very interesting, instead of going up with a skyscraper, what would happen if we dug down through these layers of cities?" said Suárez.

The Earthscraper sounds like a cool idea. But preservationists and city officials quashed it. Although, not before the design got picked up in the media. Then, the Earthscraper became a global sensation. It was a finalist in the prestigious *eVolo* magazine's annual skyscraper competition in 2010. And people from all over the world contacted Suarez to incorporate the design for their municipal plans. Variations of it were constructed. One riff on the underground concept was even built on the edge of Mexico City itself. Garden Santa Fe is a seven-story deep underground shopping mall.

Subterranean building isn't easy, which is why it isn't done very often. Plumbing has to work against gravity. Foundations have the added weight of earth to work against. Artificial lighting has to be installed to replace areas where natural light would traditionally do. Keeping spaces open, airy, and bright is tricky, not to mention expensive. All told, underground construction can be as much as five times more costly than traditional aboveground building. But Suárez says it's a must. "We need to go vertical in this city because urban sprawl cannot continue growing," he says.

Satellite towns around Mexico City have been swallowed up in what Suárez calls the "blob" of urban sprawl. The Earthscraper was a solution to try to "verticalize" in an inverse way. "This was an effort from an urban point of view to try to bring new life into the historic center and solve the problem of new living spaces, new commercial and office spaces, that you practically don't have anymore," he says.

Suárez, who turned forty at the time of this writing, was born and raised in Mexico City. He shares a passion for redesigning urban spaces with his brother, Sebastián, who is a partner in their architectural firm, Bunker Arquitectura.

The firm's name derives from Suárez's first office, which was an actual bunker in the basement of a building in the city's center. He says the office was a small space with no windows. It was all he could afford when he was starting out.

Suárez's aim is to incorporate nature into unexpected urban environments. There is a bridge, he mentions, that allows room for vegetation; a pavilion in the shape of a cactus. His designs are modern and tell a story.

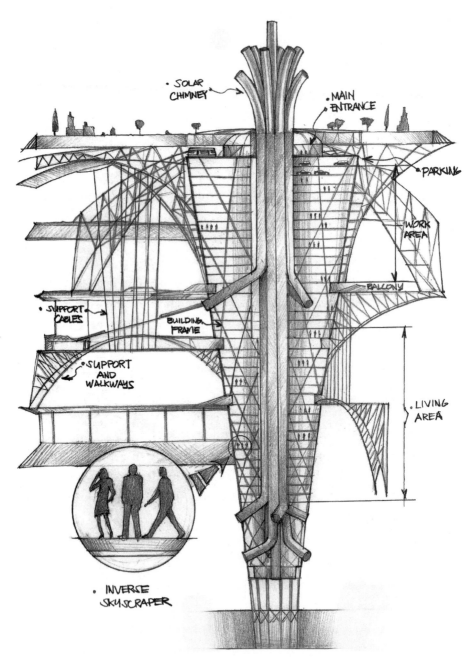

SOLAR CHIMNEY

MAIN ENTRANCE

PARKING

WORK AREA

BALCONY

SUPPORT CABLES

BUILDING FRAME

SUPPORT AND WALKWAYS

LIVING AREA

INVERSE SKYSCRAPER

The Earthscraper

He describes his architectural firm as less of a company and more of a platform to develop new strategies to study the architectural phenomenon of the twenty-first century.

In conversation, he answers questions as he might solve design problems: He quickly gets to the point without fluff. His projects are sharp, angular, pointed, yet at the same time welcoming, even if some seem upside down.

The Earthscraper is an inverted pyramid. Its design has a glass ceiling taking up nearly the entire Zócalo ground area and allowing natural light to filter down through the structure. Green-scaped walkways are lined with natural trees. A museum showcases heritage sights and Mexico's historic connections to pyramids. (Despite Egypt's association with pyramid structures, there are more ancient pyramids in Mexico and the Americas than in all the world.)

Different layers of the Earthscraper are devoted to retail, commercial, and residential spaces. Public transportation, as designed, would also pass right through the structure. Made of reinforced glass and steel, the Earthscraper looks, by its architectural drawing, modern, bright, and welcoming. It doesn't come across as cavernous, which is important. Humans fear being underground.

According to different studies, as much as 7 percent of the world's population, or about five hundred million people, are severely claustrophobic. Indeed, perceived lack of air, light, and exits brings about stress and anxiety in most people. Darkness is our biggest fear. It interrupts sleep patterns and affects people's moods. Suarez said it harkens to thoughts of being buried alive. Which is why he designed the Earthscraper to capture as much natural light as possible. The large glass ceiling would cascade light through glass floors and walls making its way to the very tip of the pyramid. There, at the bottom of the structure, a water tank would store the rainwater collected from the glass ceilings. There would be tanks for recycling water inside the building, as well as a water treatment plant. The whole place would glimmer bright.

There is something poetic about the Earthscraper rising from this point of water. It would afford its inhabitants the feeling of rising up, back to the surface; as seeds grow.

Not all underground living spaces are as aesthetically minded as Suárez's design, however. In Beijing, former bomb shelters have been re-appropriated for housing. There is no official number of how many people live down there, but some estimates claim as many as two million people live below ground.

Annette Kim, director of the Spatial Analysis Lab at the University of Southern California, spent a year in Beijing, observing and researching the lives of "rat people"—the pejorative name given to people who live underground there. She said the conditions vary, from extremely dank and squalid apartments to those not dissimilar to a basement flat in London. Humidity and mold are the biggest health hindrances, she found. "It boils down to design," she said. Those living underground in dormitory-style settings that are clean and well lit were relatively well adjusted. Those living in places designed as nuclear fallout emergency shelters fared more poorly. One woman told her that living belowground was "dehumanizing."

But China is experimenting with more pleasing designs for the future. It has ambitious plans to develop a new world underground, including plots for shopping and entertainment.

Despite China's size—the fourth-biggest in the world by land area—people need to live near centers of commerce. That's why so many choose to live underground in Beijing. They are largely migrant workers. Commuting is expensive. Living underground closer to work is cheaper and more efficient than housing farther away.

The global trend toward more concentrated centers of urban populations may force larger swaths of working-class people to accept subterranean habitats. By night, living with darkness above- and belowground may appear to be the same. For many, though, come sunrise, there will be a different type of commute, a vertical one.

Populations of zombies emerging from deep beneath the Earth's surface may be something straight out of a B movie, but increasingly, there is reason to believe we humans will be forced to spend time beneath terra firma. Heat and extreme weather will force us down.

Entire cities don't have to be built for this scene to be realized. There is a trend among the world's wealthy to have doomsday bunkers. These are residences designed to withstand the worst of man-made and natural disasters. They are stocked with supplies that can last for months or even years. They are tricked out with the latest gadgetry, some with swimming pools and screening rooms.

In San Diego, California, entrepreneur and former time-share and real estate developer Robert Vicino has made a community of these high-end shelters. They are called Vivos. One such community in South Dakota is comprised of 575 bunkers with space for ten thousand people. It is being billed as the largest survival community on Earth. And this is only one of them. There are others being built and/or planned around the world. Vivos Europa 1 is a 228,000-square-foot complex carved out of solid bedrock, under a 400-foot-tall mountain in Rothenstein, Germany.

Climate change is compelling this type of alternative housing solution. It's a rethinking of urban development, especially in tropical areas where already-high average temperatures will bump even higher. Come to Earth in a century and walk along the surface of major cities in the Middle East or equatorial areas, and things might look barren, abandoned. Beneath, though, civilizations might thrive; Earthscrapers underfoot.

Suárez's plans are ready and waiting to be constructed. And he has ideas for other buildings to accommodate a harsh future.

USC's Kim emphasizes that design is key to the health, safety, and psychological well-being of people living underground. By matching new design standards with cognitive ones, better subsurface habitats can be developed.

At the base of the flagpole in the middle of the Zócalo in Mexico City are four transparent square tiles. They cover the lights that shoot up and illuminate the giant Mexican flag that flies above. But there is a fifth tile. It's made of cement and it is padlocked. It's hard not to think that what lies beneath may be a solution to overpopulation and urban living.

A City Undersea

The Ocean Spiral is an underwater city that could be a living reality for five thousand people by the year 2030. The Japanese construction firm Shimizu has grand plans for a submerged eco-city that would be a model of how to build, operate, and sustain human life in the ocean. The project comes with a whopping twenty-six-billion-dollar price tag, but Shimizu has confidence it will get built. The plan is detailed with blueprints and site maps, as well as a definitive timeline. The construction giant is leaning on future technologies and building efficiencies, such as 3-D printing, to make it happen. Building is expected to be completely offshore and automated.

Shimizu has suggested several locations for the Ocean Spiral, based on submarine topography—from the Asian Pacific to just off the Atlantic coast of the United States. It hopes a network of cites will eventually be built deep below the ocean surface to take advantage of what it sees as the limitless potential of the deep sea. Foul weather has also been taken into account so the city will be storm-resilient.

A floating globe, 500 meters (1,640 feet) in diameter, would bob on the ocean surface. This is the top layer of the Ocean Spiral, where a glass atrium is open to the sky through clear panels. In renderings, people are shown walking along stark white floors that circle an open area. There are plants and green walls. There are diners and shoppers and kids wearing backpacks. There are flat-screen televisions and globe-like elevators. From this pleasant-looking base camp, the city spirals down in a tube structure that allows for easy transportation from top to bottom, or vice versa. Super–ballast balls are attached to the bottom of the globe to control vertical movement.

Businesses will be incubated in the middle layer. The mid-tower business zone is where companies engaged in energy, tourism, and research will locate, Shimizu hopes.

Power generation, food cultivation, and a deep-sea submarine port are housed in the layer below that. And at the bottom, 4,000 meters (about 2.5 miles) below the surface, is the Earth factory: a place for deep-sea research; cultivating and developing resources; and storing, treating, and reusing carbon dioxide.

The city is meant as a floating hub for education, experiments, and immersion in ocean-centric studies.

Ocean thermal energy will power the entire city. Food and water will also come from the sea, allowing Ocean Spiral to be self-sustaining.

"Now is the time for us to create a new connection with the deep sea because it is the Earth's final frontier," Shimizu says. "Humans will begin restoring the Earth with the power of the deep sea, using our Deep Sea Future City as a base camp."

With subterranean Earth being eyed as potential living and work spaces, subsea possibilities are sure to be explored, as well. The Ocean Spiral may be just the beginning.

Coober Pedy

Coober Pedy, Australia, located five hundred miles north of Adelaide, wasn't fancifully designed as a city for the future, it just ended up that way. And now, it's guiding others to follow in suit.

The original opal miners, who staked claims there in 1915, forged its path to global recognition, and today Coober Pedy is known as the opal capital of the world. The miners' deep-earth knowledge came in handy when they planned how to build their residences, given the punishing heat of the environs. Their homes were built out of need rather than privilege. Using their tools to dig down where it was cooler, they built their homes underground. As global temperatures rise, this is how many cities may evolve toward underground living.

Homes are dug out of sandstone and sprawl underground, interconnecting and forming an urban desert oasis shielded from the sun. The belowground structures negate the need for air conditioning. Energy until recently was diesel fuel. But Coober Pedy has turned toward renewables—solar and wind—because they are cheaper and easier to source. Trees are also being planted to provide shade. These aren't philosophical applications. They are practical solutions. The city is reacting to natural forces rather than forcing impractical design on its residents. And its practices could prove to be a model for such subterranean living

Kuwait City, one of the hottest cities on Earth, is getting so hot that speculation is it's going to be uninhabitable by 2100. That is unless city officials take leaps to cool things. And they are, but they may have to begin thinking about drilling down (for housing rather than oil). Kuwait City has learned the hard way that climate must be accounted for. Tall glass towers have been constructed over the past few decades, necessitating near-constant air conditioning. Its road system was recently engineered as a grid system, meaning transportation has become more complex and causes more pollution. Westernized aesthetics have come at a climate sacrifice.

Kuwait City isn't alone in lacking foresight. By 2050 the C40, a climate group representing the world's biggest cities, forecasts that 970 major cities around the world will be forced to rethink their relationships with heat. In a report called *For Cities, the Heat Is On*, the organization says, "The number of cities exposed to extreme temperatures will nearly triple over the next decades. . . . The urban population exposed to these high temperatures will increase by 800 percent to reach 1.6 billion by mid-century."

Redesigns will be important to stemming the displacement of millions of residents. To that end, Coober Pedy may be viewed as a city of hope. Instead of being displaced, people embraced their homes—down under.

CHAPTER 11

Stopping the Glaciers from Melting

Within this circle lies the fate of the world. Here in the Arctic, the amount of ice melt from global warming will determine the future for hundreds of millions of people. It will decide who lives and who dies, whose homes will be lost, which areas will get wiped from the surface of the Earth, and conversely where the world will blossom and thrive.

Ice melt correlates to sea level rise; sea level rise to coastal erosion and flooding; flooding to devastation. In colder areas some melt may be welcome, setting off a bloom of new plant and species life. Those places will be the fortunate few, however. Most of the world will suffer as temperatures warm and polar ice melts.

Approximately six million square miles of sea ice forms in the Arctic during winter, and about half of it melts in summertime before freezing yet again come the next winter. It is a natural glaciohydrology cycle that has been going on for millions of years. But less and less of the meltwater is freezing again. Higher ocean temperatures are causing more and more melt, and there is a decade-over-decade decline in Arctic ice extent. If this rate continues, almost all of the Arctic Ocean sea ice will melt during summers by the year 2050—maybe even a lot sooner. That will mean a clear blue ocean at the North Pole for the first time in fifty-five million years, clearing the Northwest Passage that connects the Atlantic and Pacific Oceans.

Arctic Ocean ice melt itself won't cause sea levels to rise. Sea ice is already accounted for within ocean volumes: Whether water is in ice or liquid form, it takes up the same amount of space. What happens next is what matters most and could set off a chain reaction of warming events that could put the world in jeopardy.

Warming water, as a matter of physics, expands. When that happens in oceans, coasts begin to erode, and continental ice, or the ice that has formed on land, starts to melt. This is the next step after sea ice melt and is the big danger. Glacial water on land, unlike sea ice, actually does add to ocean volumes. And there's no space for that extra volume of water to go but higher. Seas rise.

The most significant area in the northern hemisphere for sea level rise is the Greenland ice sheet. Greenland, more than three times the size of Texas, is the world's largest island. Ice covers 80 percent of it, and in some areas that ice is as much as two miles thick. All told, the ice on Greenland measures larger than the Arctic Ocean itself. If Greenland's ice were to thaw completely, it would cause global sea levels to rise twenty-four feet. And it's thawing—fast. The Arctic is warming twice as quickly as the rest of the planet.

Every foot of sea level rise equates to one hundred feet of coastal erosion. The melt puts at risk about three-quarters of a billion people who live along coastal waterways. It could cause major cities such as London and Miami to be completely swamped. Entire species could go extinct. Because the consequences are so alarming, NASA, the US space agency, has created a de facto alert system in the form of an interactive online tool. With it, people can see the effect of ice melt on their own city. The hope is they'll take steps to prepare. By clicking on their location, they can see what happens under various sea level scenarios: encroachment, flooding, or total displacement. What's frightening is that 293 major cities around the world will be affected by just a half-meter rise in sea surface. Half of a meter is a little less than twenty inches. That amount of sea-level rise is almost assured by the middle of this century and will cause more flooding, erosion, and property damage.

Ice melt will cause sea level change overall, but specific melting points combined with the Earth's rotation will target some regions before others. And not in ways that might be expected. For example, glaciers in the northeastern part of Greenland, the farthest from New York City, will affect Manhattan more than glaciers closer to it. This is because of the Earth's rotation, where the glaciers feed into the sea, and how water flows. London, on the other hand, is more closely correlated, for many of the same reasons, with Greenland's northwestern ice areas.

Sea level rise is no doubt complex. The United Nations Intergovernmental Panel on Climate Change tries to explain it simply: "It is commonly assumed that melting ice from glaciers or the Greenland and Antarctic ice sheets would cause globally uniform sea level rise, much like filling a bathtub with water. In fact, such melting results in regional variations in sea level due to a variety of processes, including changes in ocean currents, winds, the Earth's gravity field, and land height. For example, computer models that simulate these latter two processes predict a regional fall in relative sea level around the melting ice sheets, because the gravitational attraction between ice and ocean water is reduced, and the land tends to rise as the ice melts." Translation: When glacial ice melts, the land on which it has been resting tends to rise—weight is lifted.

But that simplification really doesn't do the phenomenon justice. Time, the age of ice sheets, and salinity, among other variables, also come into play. Take Stockholm, Sweden. There, the sea level is rising, but the land is rising even faster. The Earth's surface around Sweden is still making its way back up from the weight of the last ice age, which depressed the crust there as much as one thousand feet in certain places.

Given all the complexity, regionality, and time factors around surface rise, the yardstick to which the most attention should be paid is the Global Mean Sea Level, or GMSL. This is composed of all the inputs that contribute to sea level rise and provides a more uniform standard to gauge heights. Obviously the chance of the entire Greenland ice sheet melting is slim, as is the chance of the Antarctic ice sheet entirely melting (which would overshadow Greenland's melt by a factor of ten—causing more than two hundred feet of sea level rise). More sober analysis, put together by thousands of climate scientists using different modeling methods, is for GMSL to rise between six inches and six feet six inches by the end of the century. Taking the middle point of that consensus estimate would still mean a rise that would affect 145 million people. That is how many people around the globe live within three feet of sea level. Coastlines will be changed. More than a third of southern Florida will vanish. Entire countries that are low-lying, such as Bangladesh and the Netherlands, will be at major risk of flooding. This is why climate scientists are so freaked out. They aren't reacting to extreme likelihoods of sea level rise affecting the planet and masses of the human population. They are reacting to very likely assumptions.

The concern begins in a place such as this, the Svartisen glacier in Norway. Svartisen, which comprises an area of about 150 square miles, or half the size of New York City, is the second-largest glacier in Norway, sitting neatly within the Arctic Circle. The glacier cuts a form like a diamond, spilling down to a frozen point just above its fjord, the Holand Fjord. From there the water spills into the inlet that eventually opens wide to the sea.

The surrounding mountain range and landscape are magnificent—rugged, unspoiled, sparsely populated. Red cabins dot the region. White cows laze on nicely trimmed fields of green. Wild lands and forests go on for miles and miles without interruption. Winds whip down from the far north and smack you with blasts of Arctic cold air. Along the coast, the sea is fierce, and whitecaps collude in an intimidating display of strength. The seas are famously rough and unforgiving here. But it's the rock faces of the mountain range that emerge in patches of steep peaks that leave the strongest impressions. The Svartisen glacier is forged between two such rock faces that reach over six thousand feet high. The glacial ice is jagged and crinkled. Crevasses stack higher, one after another, displaying their insides and depths of blue-colored ice. The flat white surface of the glacier bounces the sun's light against the umber-colored stones facing it, washing fragments gray. Dark folds, like an old person's skin, hide in recesses on the rock. The unmistakable sound of melt—water rushing—can be heard even through the spatter of rain and winds that can eclipse fifty miles per hour. Rivers and streams are born from this glacial melt. It is life for lakes and small bodies of fresh water, power even; a hydroelectric station operates from Svartisen's flow.

Engabreen, a small section of the outlying glacier, is the lowest to sea level of any such ice formation in Europe. To the naked eye, the glacier, ninety thousand acres in size, appears solid, eternal as the bedrock on which it lies. But it is retreating at an alarming rate, shrinking to a size not seen since the Little Ice Age.

Svartisen is not alone. Glaciers the world over are receding. Ice researchers are measuring the planet's 198,000 glaciers and are finding that warmer temperatures are making more glaciers melt and melt fast. In a supersize example, many Himalayan glaciers are expected to melt away over the next century. Researchers are finding that new lakes are being formed, and meltwater is slowing. That is seriously bad news for the billon people in Asia who rely on rivers from the Himalayas to grow food, provide drinking water, as well as to power electricity. As land surface temperatures rise, lower-lying glaciers melt. The extra heat produced by that area then rises as well—we've all learned that heat rises—and then chips away at glaciers at higher elevations. Mount Everest, the tallest mountain in the Himalayas, rises 29,000 feet above sea level. And even the glaciers there are receding.

Glaciers can be found all over the world, near the poles, on mountain-tops in Africa, on islands in the Pacific. Those that contribute most to sea level rise are in Greenland and Antarctica. Glacial melt can have other ramifications in addition to sea level rise, too. The majority of the world's fresh water—approximately 70 percent—is locked up in glaciers and ice caps. When the glaciers are gone, so too is fresh water. And the cooling effect glaciers provide to the planet is important to thwart global temperature rise.

Glaciers begin as snow cover. As the snow accumulates and gets heavier, it compresses into ice. Over time, the compacted snow turns into a thicker solid form as air is squeezed out or into trapped bubbles. Eventually, gravity takes over and the ice will begin to blob—spread outward and down. Glaciers move. Slowly. Methodically. They are dynamic beasts, expanding and contracting with the seasons. Depending on the weather, they can extend or retreat. In recent decades, they have mostly shrunk.

Housed at the University of Colorado, Boulder, and aligned with a number of US government agencies to archive glacial data around the world, the National Snow and Ice Data Center (NSIDC) reports that increased global temperatures, evaporation, and wind scouring have pushed back glaciers, forcing them into retreat. Some ablation—the melting or evaporation of snow and ice—is expected. "As long as snow accumulation equals or is greater than melt and ablation, a glacier will remain in balance or even grow," the NSIDC explains. But over the past one hundred years, that hasn't occurred. Ninety percent of all the alpine glaciers that have been measured are retreating, the organization has found. "The causes of this widespread retreat are varied, but the underlying primary causes are a warming climate and the effects of increased soot and dust in areas of higher agricultural and industrial activity," it says.

In other chapters, we have discussed albedo, or the reflectiveness of a surface. Fresh snow reflects 95 percent of the sun's rays back into outer space. Water reflects just 10 percent. But soot and dust cover diminish the albedo effect of snow and ice. The reason soot and dust make such a difference is because they are dark materials and absorb the heat from sunlight, causing ice and snow to melt more quickly. The more heat that stays trapped in the atmosphere, the bigger the greenhouse effect becomes, warming global temperatures. In fact, the recent ice and glacial declines in the Earth's surface cover are like adding 25 percent more greenhouse gas emissions to the atmosphere; lost ground, as it were, that we can ill afford. To keep global temperature rise in check, the goal of reducing greenhouse gas emissions in the US alone is by more than 25 percent. The goal is to reduce emissions enough to stop any more global temperature rise.

If all the glaciers and ice caps melted, oceans would rise nearly 230 feet, inundating coasts and swamping many of the land areas of the planet. Mountain cities such as Denver would survive, but the rest of the Earth would become something out of the movie *Waterworld*. We aren't playing in the world of science fiction now, however. The chance of reducing global temperatures by cutting back on carbon emissions is just 5 percent, according to analysis published in *Nature Climate Change*. That means the melt likely won't stop. Sea ice melt is self-perpetuating: The more land and sea that is exposed after the ice is gone, the more temperatures rise and promote more ice melting. So what can we do if carbon cutbacks aren't working? A group of scientists believe they have the answer: fix the world's glaciers so they don't melt.

"We could control the collapse of the Greenland and Antarctic ice sheets," says fifty-seven-year-old John Moore, a professor of climate science and the chief scientist at the College of Global Change and Earth System Science at the Beijing Normal University in China, where he has taught for more than a decade. Moore may live in Beijing, but he travels with frequency to northern Europe's glacier areas, and is affiliated with the Arctic Centre at the University of Lapland in Finland, from which he often conducts field research. His idea is this: stall the amount of ice melt by geoengineering glaciers themselves and keeping them frozen in order to buy time to fix climate change. If by the end of the century most coastal cities will be threatened by sea level rise, as expected, then why not figure ways to put off that risk until society can effectively deal with its carbon emissions problem? The economic costs alone of sea level rise are too great to wait, he says. Moore calculates that without coastal protections, the global costs of sea level rise will be fifty trillion dollars per year by 2100. He claims that glacial geoengineering solutions would cost far less and provide immediate results.

Moore is able to cite numerous examples from different locations across the globe where glacier tests have worked: in Asia, in Antarctica, in Switzerland, and, of course, in the Arctic region. Rupert Gladstone, a geoscientist at the Arctic Centre; Thomas Zwinger, a senior application scientist at the CSC-IT Center for Science in Finland; and Michael Wolovick, a glaciologist at Princeton University, have teamed with him to work on three unique ways to stall the fastest flows of ice into the ocean. First is a plan to stop warm water from reaching glaciers. By blocking warmer water from reaching ice, melting would slow, more ice would form, and icebergs would lodge. This would be done by dredging along continental shelves to create artificial embankments. "This artificial embankment, or berm, could be clad in concrete to stop it being eroded. The scale of the berm would be comparable with large civil engineering projects," the scientists say, citing the Suez Canal, Hong Kong's airport, and the Three Gorges Dam as examples of projects that required large-scale construction and a huge amount of materials, more materials than would be needed to build a concrete sea structure.

Another solution would be to artificially prop up ocean shelves. This would be accomplished by pinning the ice shelves in front of glaciers to the seafloor. By erecting man-made islands, warm water would be blocked from encroaching on the glacier. Their third idea is to dry out subglacial streams. "Fast-sliding ice streams supply 90 percent of ice entering the sea. As the ice slides over the glacier bed, frictional heat generates . . . at the base of the ice streams. This water acts as a lubricant, speeding up the flow, which in turn generates more heat, and creates more water and slip-page," Moore and his colleagues explain. By draining the stream, the flow slows down and allows enough time for the ice to thicken.

This last idea of slowing down flows seems the most reasonable. There is precedent for it at a couple of places already—at the Svartisen glacier and in the South Pole.

At Svartisen, glaciologists have drilled tunnels in the bedrock under-neath the glacier to drain it. The drained water flow feeds the hydro-power plant and at the same time provides researchers like Moore with a live example of how tunneling elsewhere might work to preserve gla-cial ice.

Inside the Engabreen tunnel at Svartisen, a full laboratory has been designed to monitor the ice. Beds, a kitchen, a bathroom, and workspace allow scientists to spend a significant amount of time studying how wa-ter moves within the glacier and the composition of ice that is formed. Some call it the most claustrophobic lab in the world, but it is an impor-tant outpost for developing models that could show how to stop ice sheets from quickly disappearing.

The trek for a peek at the tunnel isn't easy. The journey begins with a boat ride across a lake. The sleek black-and-gray, open-deck, flat-bottom boat looks like something out of a James Bond film, and the captain races it as if he is chasing villains. You stand, with only a sort of saddle between your legs, and strap into a body harness and life preserver. After just ten or so minutes blasting along the lake, you arrive at a quiet, bu-colic haven. The boat docks in front of a red boathouse from another century, and the walk begins, from woods around the lake to the glacier in the distance. There is a small farm, a cowherd, a moose on display. Farther afoot is a snack house and observation deck on the shore of the fjord. The last boat departs at 7:30 P.M., and then you are on your own for the night.

The trek to the glacier begins in earnest at the fjord. Two marked paths go up the mountain toward the glacier face: A short path leads to a full frontal view of the ice, and another banks left and goes higher up into the patch of trees near the mountain summit. That's where the tunnel lies, burrowing down 650 feet underneath the surface.

Red marks guide you along rock faces, cracks, and streams. There is a chain secured by steel rods that you can grab onto. It's easy to slip, or fall, over the steep slope. The chain ends at the bottom of the glacier, and from there it's all balance and wits. It gets steep as you climb above the tree line on slippery mud, moss, and rocks. The red trail marks play tricks with you, forcing you to boulder and wind your way up. When the Arctic weather sets in—in a flash—dark clouds, rain, and heavy wind add to the trek's challenge. Air becomes precious and you gulp it, heaving as you go. Your thighs burn for the vertical steps you are taking. It's a slow pace. You notice the silence, the rush of water down a stream, your breath. You fight the wind. The other elements around you—the trees, the grass, the rocks, the ice—have long ago learned to adapt, give way, or tumble to another place to settle. It's a reminder that Earth was here long before humans and will be here long after we are gone. Our species' survival rests on figuring how to adapt, to march on.

Inside the tunnel's laboratory, scientists are figuring ways to do just that. They are literally calculating history from ice core samples and inventing solutions for the future. By inserting themselves inside the glacier, sometimes for months on end, they can better understand its inner workings. That knowledge might allow us to reengineer masses of earth and ice that took millions of years to form.

The Engabreen tunnel, for example, helps with Moore's research, and that research, in turn, has found major support. The Chinese government is investing three billion dollars into polar research, some of which is going toward Moore's geoengineering project. Besides the environmental effects that the clearing of the Northwest Passage poses for the world, the passage's clearing also has major economic implications for

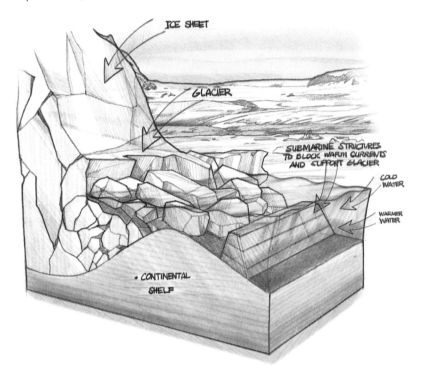

ICE SHEET

GLACIER

SUBMARINE STRUCTURES
TO BLOCK WARM CURRENTS
AND SUPPORT GLACIER

COLD
WATER

WARMER
WATER

CONTINENTAL
SHELF

The Glacier Stanchion

mining, fishing, and trade. (Shipping companies can cut thousands of miles off routes because the Atlantic and Pacific Oceans will be more easily connected. Ships will be able to travel north instead of using the long route south around Africa.)

No matter what the overall goal of the funding is, China's support of Moore's research could make geoengineering glaciers happen more quickly. "We've already doing simulations," he explains. While actual glacial fieldwork has begun, it will require some years of observation before any construction would begin. Each proffered solution for glacier melt would start on a small scale as a test, Moore says, and depending on the results, efforts would be ramped up from there. Results mean more than stopping the melt. How the construction and manipulation of glaciers affect local habitats will also be observed and taken into account.

Moore and his fellow ice invaders are well aware of the hazards they could bring about. For each solution they have charted, they also discuss the dangerous working conditions and environmental perils. Redirecting warm water by building berms, for example, means "construction would be arduous and potentially hazardous in cold waters littered with icebergs." It also means the local marine ecosystems would be affected in unknown ways by all the turbulence and sediment disruption brought about by the construction. Long-term, further losses would likely accrue once the glacial melt is slowed. Ocean environments, sensitive to temperature changes, would be altered by a change in ocean mix, which affects entire species and life patterns.

Erecting artificial islands or draining glacial streams would bring about similar negative consequences. But these scientists remind us that the greatest risk is doing nothing. "The impacts of construction would be dwarfed locally by the effects of the ice sheet's collapse, and globally by rapid sea level rise," they say.

A future world, then, might have the whole of the Arctic Ocean traversable—wild and blue. A new map to be drawn for the planet. At critical glacial junctures off Greenland or in the Antarctic there would be embankments, strings of island barricades, or pumping stations. These formations would likely distribute new kinds of ocean oscillations, and produce an alternative reality of weather and seasons. How weather will change and where is for the most part unknown.

Meteorologists are putting together computer models to account for what will no doubt be a change in sea surface temperatures when the Arctic melts away. They already know that a warmer north weakens the polar jet stream, sending colder air farther south. (Evidenced by the famous polar vortex North America gets during winter.) A chronically warmer polar jet stream could trigger a whole set of other weather sequences. For sure, weather extremes will be heightened. If the Arctic melts, "heat waves, rainfall, drought, blizzards and hurricanes may all become more common," according to a report on

AccuWeather.com, the leading source of weather information for the planet.

Weather, sea level rise, flooding, population displacement; geoengineering glaciers could stop the imminent flow.

Ice911

Ice911 has a catchy name, and the nonprofit organization's mission is even more provocative: lowering global temperatures by restoring ice in the Arctic. It aims to spread fake snow (sand-like material made from glass) across the Arctic Circle to reflect the sun's energy and keep the surface cold.

The reflective material mimics the effect of bright ice, bouncing back into the atmosphere 90 percent of the sun's heat that hits the ground. The material is made from silica. Ice911 claims that it's harmless to humans, animals, and local ecosystems. The nonprofit notes that silica can be found in many foods and food products, and is often used for animal feed as well. "Best of all, the silica microspheres slowly degrade," Ice911 says.

Dr. Leslie Field, a Stanford University professor, invented the material and founded the organization after ten years of research and testing.

In 2017, the reflective granules were spread across four acres of ice in the Alaskan Arctic. The test proved positive and the goal is to within a few years deploy four times that amount.

According to the group's preliminary climate models, spreading the material on ice in the Arctic can reduce average temperatures there by 2.7°F; reduce overall global temperature rise; increase ice volume in the Arctic by 10 percent within forty years; and increase ice thickness.

Ice911 is trying to preserve a particular kind of ice: multiyear ice. This is the most reflective type in the Arctic and can stay frozen through summers. But in recent times, multiyear ice has been melting quickly. By covering vast areas, the materials, which also float, can protect ice surfaces, preventing melt and the maladies that go along with it, such as sea level rise.

The material is inexpensive, costing just one cent per 10 square feet. But the Arctic Circle is 108 square miles in size, or millions and millions of square feet, making the project rather pricey to be put into practice. Although Ice911 isn't planning to cover the entire Arctic Circle, to cover any significant area would amount to hundreds of millions of dollars.

The reflective sand-snow could be spread by ship or plane, or by other means on the ground. Skeptics have raised concerns that the materials may not be as benign or eco-friendly as suggested. Critics believe weather patterns and other unforeseen consequences could accrue by artificially interfering with Arctic sea temperatures.

Ice911, though, remains resolved to fixing the problem of Arctic sea ice melting.

Ice Stupas

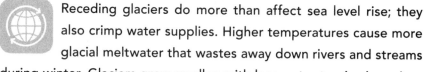Receding glaciers do more than affect sea level rise; they also crimp water supplies. Higher temperatures cause more glacial meltwater that wastes away down rivers and streams during winter. Glaciers grow smaller, with less water to give in springtime, during the traditional melt season. The Ice Stupa Project is a way to delay the glacial melt, saving it for when water is in higher demand for agriculture.

Sonam Wangchuk, a fifty-two-year-old engineer in Ladakh, a sprawling region in northern India known as the Trans-Himalayas, devised the solution. By erecting a vertical pipe fed by water flowing downstream, mist sprays out the top. The winter cold freezes the droplets into a cone shape and that provides shade for the ice layers that form

The Ice Stupa Project

underneath. The cone shape keeps the ice frozen longer. When the ice stupas do melt, it provides a ready source of water and, in turn, irrigation for rural villages in the region.

A stupa is a sacred mud structure in Ladakh. And the artificial glaciers that Wangchuk created look similar to them. They look like giant melted candles. No machinery is needed to construct the ice stupas—just a pipe and hard work.

The Ice Stupa Project proposes to build dozens of artificial glaciers that will allow trees and vegetation to be planted in the high desert region—in many places for the first time. The ice stupas will change the mountain landscape.

An institute associated with the project is also being planned to teach students about mountain development and climate change adaptation. The hope is to develop scientific solutions and technologies for the unique problems faced in the high mountains.

In 2016, Wangchuk won a prestigious Rolex Award for Enterprise, and he has launched a campaign to increase the number of artificial glaciers built every year.

The Ice Stupa Project may not be able to stop glacial ice from receding, but it can delay when the water flows, and redirect it to help prevent some of the more worrisome effects caused by glacial melt.

HUMANS V. NATURAL RESOURCES

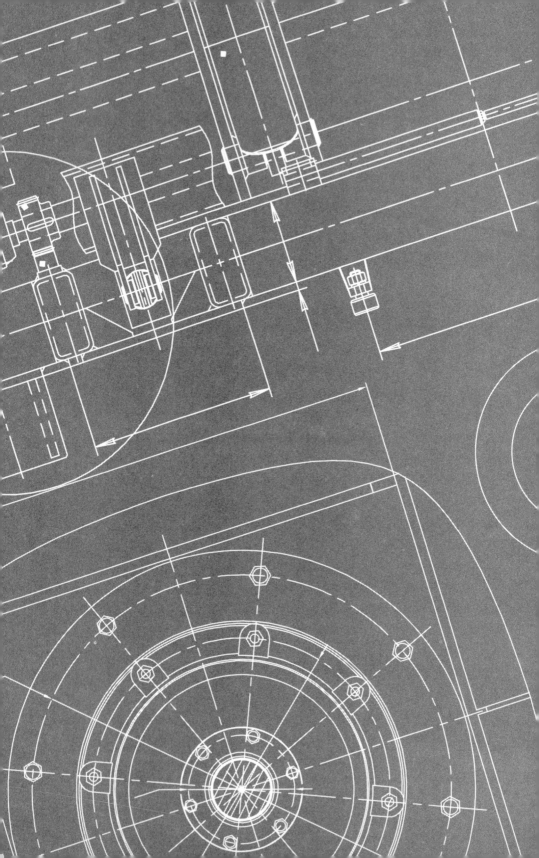

CHAPTER 12

What Lies Beneath

 It is a tragic irony. Brazil is the most water-rich country in the world—with 20 percent of the planet's fresh water resources— yet it is suffering from thirst.

Brazil's famous rainforest, the Amazon, is about half the size of all of Europe combined and, as is implied, sees heavy rainfall. The regular rains add to rivers and basins. The jungle laps up water and feeds its community: the plants, trees, and animals, including us humans, who live on all that the rainforest produces. Timber and livestock and vegetables and oils and spices and even the ingredients for many of our medicines come from the species-rich habitat.

This amazing ecosystem, a monumental wonder of nature, covers about half the country of Brazil. But access to the fresh water the rainforest provides is fragmented. Energy and agriculture are terrific consumers: Nearly two-thirds of Brazil's energy comes from hydropower. And irrigation consumes nearly three-quarters of the total freshwater supply. Add climate change to depletion rates, and what remains for human consumption and sanitation is difficult to manage.

Despite being the world's fifth-largest nation by size, with two hundred million people living in it, and a treasure trove of natural resources, Brazil isn't a wealthy country. The average annual income per capita is a bit more than fifteen thousand US dollars. The national debt is huge.

And corruption has run rampant at the highest levels of government. Locally, problems range from poor infrastructure to business inefficiencies, which makes building a world-class water delivery system a daring moonshot. But Brazil is going for it.

Over the past few decades, Brazil invested close to four hundred billion dollars in infrastructure projects throughout the country. It has created the largest market for public-private partnerships in Latin America in hopes of creating more private sector opportunities, such as jobs. And at the core of its master plan is the sprawling water delivery system, a geoengineering project that will transport water from regions rich in water resources to those where access is chronically difficult. Once completed as projected, Cuncas will become the world's longest water tunnel, stretching three hundred miles. It has been called the eighth wonder of the world and likened to the building of the Egyptian pyramids. However, bureaucracy and ballooning budgets have stalled its completion.

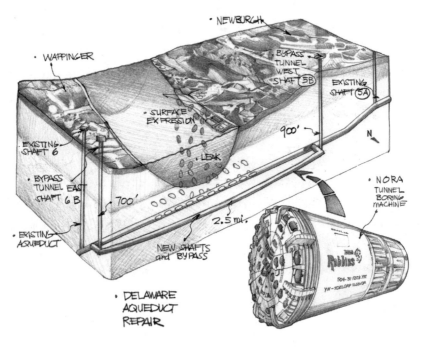

The World's Longest Tunnel

Until Cuncas is totally completed or a grander subterranean project comes along (there are plans in China for water tunnels that could exceed the length of Brazil's), the honor of being the world's longest water tunnel goes to the Delaware Aqueduct, which lies deep beneath New York City.

The Delaware Aqueduct is eighty-five miles long and runs from upstate, north of the Catskill Mountains, down into the city's boroughs. It was built, almost unbelievably, over the course of only five years, beginning in 1939.

The waterway carries half a billion gallons of water a day from countryside reservoirs to city high-rise taps. It was made by blasting, drilling, and boring through hard rock and deep earth—a phenomenal achievement, especially for its time.

The aqueduct dives as much as 1,500 feet below the Earth's surface and winds through small towns, under roads, and beneath rivers. The tunnel itself measures between 13.5 feet and 19.5 feet in diameter and is lined with concrete.

But like many water systems around the world, New York's is in a dismal state of disrepair, hemorrhaging up to thirty-six million gallons of water a day for at least the past twenty-five years. (You read that correctly.) It sounds like a crazy amount, until you look at the global statistics. Globally, forty-five billion liters (twelve billion gallons) are lost per day, mostly due to water leakage. In developing countries, almost half their fresh water never makes it from sources to people because of poor pipes and plumbing. The global number put on all this loss is fourteen billion dollars a year.

It's worth restating World Health Organization statistics: More than two billion people lack safe drinking water at home; more than eight hundred million people do not have ready access at all to fresh water. Water cannot, and should not, be wasted.

Eyeing a water-scarce future, New York is investing two billion dollars to upgrade and fix its leaky aqueduct. The repair job, begun in 2013 and scheduled for completion in 2022, involves more boring, tunneling, excavating, and digging—but utilizing twenty-first-century technologies such as an auto-piloted submarine (technically an "autonomous underwater vehicle") and the world's largest boring machine, named Nora (in honor of the first woman, Nora Stanton Blatch Barney, to receive a college degree in civil engineering).

A new tunnel extension has to be added to the system to fix the leak. And during its final stage, the whole thing will shut; New York will be without its main supply of water. Backup supplies are promised to fill the gap. They had better work. New York City without enough water for people and businesses would likely erupt into mayhem.

Human development of water flows, it seems, doesn't come without major difficulties.

The Cuncas tunnel project in Brazil is learning this the hard way. It has been straddled with repairs, and industry reports say what has been built may have to be built again. The going hasn't been easy, admits Pádua Andrade, Brazil's minister of National Integration and the former secretary of Water Infrastructure, which is responsible for the São Francisco project. He says difficulties mount due to the complexity and the size of the structure.

For the project to succeed, water must be captured at the São Francisco River, a major, eighteen-hundred-mile-long system within Brazil, and transported through new tunnels and over remote hills to a vast network of canals, aqueducts, and reservoirs. The water will service the rural communities in the northeast part of the country and prop up farming communities, as well as major cities in need. The audacious plan was put into effect by one of Brazil's most strong-willed political leaders, President Luiz Inácio Lula da Silva, who was born in the water-starved northeast region.

The northeast is drought prone, and the São Francisco River, which begins in the southwest and banks out to the coast before reaching that far north. Along the way, coastal cities such São Paulo, which has come within twenty days of completely running dry, are bypassed, so the new tunnel source should be welcomed by twelve million urban dwellers as well as rural farmers.

Pretty much anywhere on Earth water travels quite a distance before it comes out of a tap. And mechanically, most water systems operate similarly. Rivers connect to aqueducts, where water is treated to ensure it is potable. This treatment typically involves a number of steps: screening, coagulation, flocculation, sedimentation, filtration, and disinfection, among other processes. Fluorides are usually added to help tooth decay. Carbon is sometimes added for taste. And there are corrosion-control additives for pipes.

From the treatment plant, water flows to a storage tank and then through large pipes to water mains in streets. Plumbing from homes and commercial buildings connects to mains, and then water is ready to pour out of taps.

In the developing world, in places such as Brazil, this maze of water engineering is still being assembled. Without it, the choice is largely to walk for water. Where water is scarce, people (mostly women) walk an average of 6 kilometers (about 3.75 miles) per day to retrieve water from ponds, lakes, wells, streams, or what have you. Try that with a forty-pound can on your head, as is custom for about a quarter of a billion people in developing countries around the globe.

In many parts of Brazil, tube wells are used to siphon water from wells or underground sources. A tube well is a simple construct where a tube or pipe is dug underground and connected to an underground water source. A pump is fitted to the top to siphon the water up and out. Of course, tube wells skip many steps in the filtration process. They are primitive. The only way to deliver vast amounts of water to millions of people is to funnel it in large-scale amounts. Nature does this via rivers, lakes, streams, and underground aquifers. Natural underground aquifers are filled by rain or snow and ice melt seeping into the ground, or by river wash.

The largest underground water system was discovered in 2018. A 215-mile-long flooded cave was discovered in Mexico. It connects the world's largest flooded caverns—Sac Actun and Dos Ojos in the Yucatán Peninsula. The discovery eclipses the 167-mile-long Ox Bel Ha system that is also located in the Yucatán, just south of Tulum.

The Delaware Aqueduct and the Cuncas tunnel projects are essentially trying to mimic these caverns: nature's unseen water feeding systems. But replicating megalong caverns is expensive and requires ongoing diligence. Nora the boring machine, for example, works five days a week, twenty-four hours a day. When her work is done, billions of gallons of water will flow through the tunnel she has dug. On the flip side, it would take nature thousands of years to do that.

When you are trekking through the Amazon rainforest in Brazil, water scarcity is hardly what comes to mind. There are moist leaves underfoot, streams and rivers to cross. The rains come with regularity. Wading through hip-deep waters, you are likely more concerned about alligator-like caimans than anything else. You are actually hoping for less water, not more.

You usually wake to light mists and dew coverings. Water appears in abundance—everywhere. Just chop a vine from a tree and drink from it. Walk to a depression in the forest and there is likely a water hole. Streams, some fast moving, some a trickle, are hardly ever out of sight. Fetching water any other place on the planet is rarely this easy.

Getting water from one place to another is a haunting problem for humanity. Some countries have even turned to importing water by shipping tanker. In others, they dig deep for wells. In the Great Plains of the United States, farmers use pumps to drill hundreds of feet down to tap the Ogallala Aquifer, one of the largest underground sources of fresh water in the world. (Which, as a side note, is drying up, according to geologists. But that is a story for another time.)

So boring massive tunnels is really the only way to move large amounts of water consistently from one area to another without huge energy costs. In California, for instance, transporting water is the state's biggest single consumer of energy, accounting for 20 percent of the government's electricity budget alone, never mind the gas and diesel utilized.

Power is needed to supply, treat, distribute, and recycle water. It's needed for pumps, processors, and motors; vehicles, too. Tunnels obviously do not require continuous electricity to make water move. In fact, an advanced water system such as Cuncas's will create power.

We typically associate dams with hydropower. But dams are nothing more than grand man-made water diverters and are typically baked into major water-systems projects such as the Cuncas tunnel and the Delaware Aqueduct.

Damming has come a long way since it was begun in the twenty-ninth century B.C. by the ancient Egyptians. Those dams, basic gravity dams, were made of bricks, and their only function was to block water. Damming today could come with hydroelectric power, flood control, and technologies that automatically manage flow and discharges, depending on weather conditions and demands.

There are at least a dozen common types of dams (arch dams, buttress dams, coffer dams, diversion dams, embankment dams, gravity dams, hydropower dams, industrial waste dams, masonry dams, overflow dams, afterbay dams, dikes). And they come in all sizes—thousands and thousands of them all over the world. Fifty-seven thousand of those dams rise more than four stories tall. These are the big ones we often see on television or in movies: huge arch dams such as Nevada's Hoover Dam or Switzerland's Contra Dam (famously featured in the James Bond film *GoldenEye*). The biggest dam in the world is the Three Gorges Dam in China. It is an immense structure that captures water in central China from the Yangtze River, which is the longest in Asia.

The Three Gorges Dam is approximately 600 feet tall, 1.4 miles long, and 130 feet wide. It creates a virtual lake more than 400 miles long, doubling the size of the natural river channel that was there before it was built. Construction began in 1994 and the dam began operating in 2003. It produces more electricity than any other power plant in the world and services millions of homes. Some reports say it alone meets 10 percent of China's energy demand.

By creating dams, we humans took a huge leap in fashioning nature to our liking. Although it has come with consequences. Dams have a widespread effect on species endangerment, loss of forests and agricultural lands, flooding, sediment loss, land infertility, stream flow, and lower oxygen levels in river waters. Dams throw ecosystems out of whack.

There is an international antidam movement made up of thousands of environmental, human rights, and activists groups. They are looking to stop the construction of more large dams. One of their more famous members is Yvon Chouinard, the billionaire founder of Patagonia, the outdoor apparel company. "The fact is, dams are dirty—and their destructive impact far outweighs their usefulness. In particular, the electricity they generate can now be gained much more effectively from other sources that don't disrupt rivers, destroy habitats and displace people. If these misguided projects aren't stopped, the ecological damage and impact on local communities will be devastating," he wrote in a blog post for Patagonia's website.

Critics say the Cuncas diversion will further deplete the São Francisco River, already suffering from lower levels because of climate change and overconsumption, reducing its flow and the amount of water available to people who already rely on it. They fear the river will totally dry up if damming and diverting continues. But to service growth, dry regions need water. Pádua Andrade, the Brazilian government minister who is also an engineer, notes that it is the water that is diverted, not the river itself. "There is no diversion of the river as its natural course is not changed. It is an integration of existing basins," he specifies. Which seems semantical until you consider the rerouting of the Mississippi River here in the US. We've blocked its natural path and forced it in the directions we humans prefer. That has caused all sorts of consequences, including massive floods. At the São Francisco River, even if the river isn't forcefully bent in a new direction, the water flow will surely lessen due to the dispersion.

In the late 1920s, New York City's water department decided it had to do something as big and as grand as nature would to feed its growing population. It had tapped as much water as it could from the surrounding area, impounding water from the Croton River in Westchester County and directing it to Manhattan and the boroughs. But even more water was needed. After careful analysis, officials, according to the state record, decided that the watersheds of the Catskill Mountains, just a couple of hours north of Manhattan, would provide the most reliable resource. Problematically, the Catskill watershed also services the state of New Jersey. A lawsuit was settled by the US Supreme Court, allowing the New York City diversion, and construction began a decade later. For nearly a hundred years the aqueduct has held the title as the world's longest man-made tunnel—an unsung marvel of service.

In 2001, the terrorist attacks on the World Trade Center shined a new light on the water system. Fears mounted that the supply would be sabotaged. Access to the Rondout Reservoir, for example, where the aqueduct draws much of its reserves, has been limited and security cameras installed. It's off limits. But you can visit the Ashokan Reservoir, a sister reserve in the Catskills that also services New York City. It relies on more clandestine security-means to keep the water safe. It also tricks nature.

Standing on its shore and looking across the basin on a bright autumn day you'd never guess the foliage that accompanies the three million mature trees at the forest's edge are artificially planted there to condition the soil and keep the watershed sound. You'd never guess that there are steel gates underneath the Dividing Weir Bridge that can plunge 180 feet to lock out water as a way to protect sediment flow. Or that there is an entire police force, engineering group, and monitoring protocols in place to keep watch over the water. Water executives, officials, and foreign dignitaries all visit to find out how New York City manages to keep the water so fresh at its raw source.

New York City water isn't filtered. It is one of only five large municipalities in the country to deliver unfiltered drinking water to its residents.

Fresh off a morning hike, Adam Bosch, director of public affairs for the New York City Department of Environmental Protection, points to a mountain peak high in the distance: Slide Mountain. At 4,204 feet, it is the tallest point in the Catskill Mountains, and it's where, Bosch says, New York City's water journey begins.

Bosch is only in his late thirties, but he has an amazing command of the history and facts of the centuries-old water-supply system. He can tick off the engineering behind how water is delivered as easily as he can offer an interesting tale or tidbit related to the system. (Back in the day, workers dynamiting tunnels used to smoke and drink inside the caverns where the dynamite was stored; trains used to lock their toilets when traveling over bridges in the area to ensure no effluent got into the reservoirs; and then there is the story of Carl, the bear whisperer, employed by the city to educate Hasidic communities who frequent the area about recycling and proper garbage disposal to keep bears away, not to mention keep the water clean.)

One particularly fascinating nugget of scientific insight is how bluegill fish are used to test for water quality. If the fish begin to show signs of abnormal behavior, there is likely something afoul with the water. Their natural signals have proven to be more sensitive to toxic intrusions than the most sophisticated water-testing devices. Once, they detected a tiny fuel-oil spill—something not picked up by lab instruments because the spill was so small.

The accompanying stories Bosch recounts are nearly as interesting as the science of transporting fresh water from here to New York City, more than one hundred miles away.

From the peak of Slide Mountain, water (meltwater and rainwater) makes its way down thousands of feet to the Ashokan Reservoir, which is at an elevation of 590 feet. From there, the water continues farther down, flushed with the power of gravity, through Ulster County, where it intersects with the Delaware Aqueduct. Just south of the town of New-burgh, the water flows under the Hudson River and pops out in Putnam County, where it weaves and arcs and zigzags along to the New Croton Reservoir, at an elevation of 196 feet. Engineers clearly maximized the use of the 400-foot differential between there and the Ashokan Reservoir to keep the water moving. Next, the flow banks along through Westchester County and the Bronx, where it makes its final push to Manhattan. Here, at just 15 feet above sea level in some places, it trickles, streams, and drips out of taps.

The route is a fascinating use of gravity, and it showcases nature's clean power. Even at its endpoint within the city's borders, enough energy is left over from the water's journey to push it up four or five stories. That is why there are so many four-story walk-up apartments in New York City; a hundred years ago, water could service apartment units just fine without the need of man-made energy.

On its way down, the Ashokan water must travel through steep plateaus and rural, wild areas. Its biggest obstacle, notably, is the Hudson River, under which it must travel through tunnels more than a thousand feet deep.

Brazil's Cuncas tunnel faces steeper logistical challenges. It must forge through lands not predisposed to carrying water. That means soil reacts differently and can cave. It must bore through hills, bringing about support and buttress issues. It must travel aboveground and belowground, necessitating different infrastructures. And it has to hold up, meaning building materials and engineering efficiencies must be—if the tunnel is to last for centuries—of the highest quality. That has not always been the case. Sections are already being rebuilt, as mentioned; delays are causing patches to crumble.

As with all underground excavation works, says Andrade—the minister who has overseen the project from one aspect or another from its beginning—in spite of the testing and modeling that precede construction, each excavated meter can change once ground is broken. "It's the complexity of executing structures of this size," he says, that slows the tunneling down.

The first pumping station for the Cuncas tunnel is in a remote corner of Brazil in a town called Cabrobó. It's where the São Francisco River turns toward the sea, depriving the northern states of Ceará, Paraíba, Rio Grande do Norte, and Pernambuco of its fresh water.

The first section of the tunnel will go from Cabrobó, which is in the state of Pernambuco, to a reservoir 5.5 miles away. It will travel through scrub brush and mud, along embankments and by roads. Other sections of the tunnel will splinter into aqueducts and canals to deliver water to the other dry states.

According to Jacqueline Rocha, the spokesperson for the government agency that oversees the tunnel's construction and management, things are, despite reports to the contrary, on track to get the water permanently flowing. It will have taken more than a decade.

This is why New York's aqueduct, completed in less than half the time, is a world-class model that still confounds. It proves that purpose trumps progress. Tunnel workers and managers alike believed they were on a historic public-works mission that would stand the test of time. It's why they wrote poems about their labors and etched intricate designs into tunnel hatches, doors, and gates. They made it a true monument.

Standing on the New Jersey side of the Hudson River and looking across at the storied skyline of Manhattan, one sees the jagged lines of tall buildings that sketch a rough silhouette at dusk. It's hard to contemplate swimming across, never mind tunneling under the river, to reach that image. But there is a virtual city underground, where workers at any given time, six hundred feet under the surface, are prepping the space for the aqueduct's transfer; its shutoff for the first time in its history.

These are the new lengths we must go to in order to support our modern-day settlements. It's obvious just by looking at New York City from afar that so many millions of people shouldn't really be living together on one relatively small island. But we do.

More people living in urban areas means more people will need water in places that are already stretched for supplies. The water will have to come from other sources, likely farther away. Yet, more urbanites doesn't mean that existing centers of population will be the only ones added to; new metropolitan areas are being developed. And these new cities and towns will also need water delivered.

In Asia, new cities are being planned in expectation of grand water-delivery systems. In countries like Brazil, new water routes will mean a shift in traditional centers of trade and commerce. Topographic and political maps may have to be redrawn.

Cities have historically been built near or around bodies of water. Innovation dispenses with that notion.

In the future, rivers will bend to our needs. An unnatural, subterranean world may become an unseen water factory—beltways of resources belowground moving H_2O to the masses. What lies beneath could solve the mystery of how cities will function, and how water will make its appearance in places where it has gone missing.

China's Megatunnel

 China is in the process of developing the world's longest tunnel, which would transport water through some of the planet's harshest, most arid terrain. This six-hundred-mile-long water passageway will deliver the much-needed resource from a river in Tibet, high in the Himalayas, to the Taklamakan Desert floor in northwest China.

More than one hundred scientists are reportedly working on the project, which is designed to divert as much as fifteen billion tons of water per year to make the desert bloom. That's about 25 percent as much water as flows through all of China's Yellow River, a terrific amount. The goal is to complete the megatunnel within a decade.

Given a steady source of irrigation, the Taklamakan Desert could prove to be a formidable agricultural zone, as opposed to the wasteland it is today.

The massive undertaking is purported to cost approximately nineteen million dollars per mile. Advanced engineering techniques are being utilized so the tunnel can more easily pass through fault zones. The tunnel has to drop thousands of feet and snake through steep mountain gorges to reach the desert floor. Waterfalls will cascade water from one tunnel section to another. Large concrete pipes bound by flexible materials have to be hoisted and connected. There are boring, clearing, and enough logistical issues to make the project a construction nightmare. Until now, the altitude and watery depths to consider have kept the artificial waterway out of reach.

Tibet's Yarlung Tsangpo River has long been eyed as a water resource for China and there have been impossible plans for centuries to tap it (tons of water dropping thousands of feet crushes pretty much anything in its path and heretofore couldn't be properly managed). Nowadays, innovation, computer modeling, and new equipment make the tunnel possible—well, feasible. Plans are still in their blueprint stages.

The Yarlung Tsangpo River is fed by melt from the Himalaya mountains and monsoon rains. In turn, it feeds myriad other rivers and tributaries, including the mighty Ganges in India. The water diversions to other countries make the tunnel project very controversial politically. Tibet, while technically part of China, is an autonomous region and its sovereignty has been a long-standing matter of debate and conflict. Additionally, diverting headwaters will no doubt strain Sino-Indian relations. Also looming large are geological challenges: The tunnel will transform a mostly uninhabitable area about the size of Germany into a fertile valley that could be developed and populated.

To prove the tunnel's possibility, China is building yet another long tunnel, in its Yunnan province, in the southwest part of the country. It is also an arid region, which will source water from a towering plateau.

Hundreds of large-scale water projects throughout the country have been embarked upon to support the government's ambitious plans for growth. By remapping its geography, China can harness more productivity and, in turn, more economic power.

Terraforming the Taklamakan Desert would be quite a feat. The engineering accomplishment of the tunnel construction is one thing, but the Taklamakan is the largest desert in China and the biggest "sand sea" in the world. Converting such a vast area of sand to workable soil would change where and how nature decided things should grow. Or not.

CHAPTER 13

Food Animals

There are 7.7 billion humans and counting on Planet Earth. Of all, the greatest number of starving people reside in Africa, south of the Sahara, in the countries Ethiopia, Somalia, Eritrea, Djibouti, Kenya, Sudan, and Uganda. The region, known as the Horn of Africa, is a pointy piece of the northeast corner of the African continent, across the Gulf of Aden from Yemen, with a population of 160 million.

Food insecurity varies, of course, by place and time, even outside of the Horn. One year a drought could cause famine, as it did in South Sudan in 2017. Another year a flood could decimate croplands and food stocks, as it did in Malawi in 2015.

Conflict, too, in different countries can take food out of the mouths of many. In Yemen, the civil war begun in 2015 has caused the world's worst food crisis. Seventeen million Yemenis have been dragged into starvation by modern times' worst humanitarian crisis. Every ten minutes a child under the age of five dies there. On average over the course of decades, however, the Horn of Africa has suffered from malnutrition more than anywhere else. Jeffrey Sachs, director of the Earth Institute at Columbia University and a renowned economist, has called this 727,000-square-mile region the most vulnerable in the world.

Even in a big city like Addis Ababa—the capital of Ethiopia which houses United Nations representatives, numerous international agencies, and has modern facilities and advanced communications—people go hungry. Farther out in the countryside, in rural communities, conditions worsen. Subsistence cultures rely on the land for their diets, and often the land refuses to cooperate. Ethiopia experiences chronic weather extremes that corrupt agricultural conditions and cast its population into serial despair.

Although hunger rates are bad, they are better than they have been. In 2000, the child malnutrition rate in Ethiopia was 58 percent. Six years later that number dropped significantly, to a 38 percent rate of child malnourishment. Globally, overall hunger has declined since the year 2000. The Global Hunger Index, produced every year by the International Food Policy Research Institute, shows that the level of hunger in the world decreased by nearly a third over the past two decades.

Technology, education, government policies, advances in communication via mobile phones, and investments in small-scale farmers have lifted masses out of malnourishment. Still, there are almost a billion people around the world who tuck in to bed hungry each night. By 2030, the United Nations has pledged to help each and every one of those people and eradicate world hunger. It's a lofty target. Some say it is unachievable. But significant progress, as noted, has been made over the years. In total, the number of malnourished people in the developing world was halved over a twenty-five-year period, from 1990 through 2015. And fewer people over that time reported living in extreme poverty, on less than $1.25 a day.

For the hungry, however, numbers and global trends aren't comforting. In rural Ethiopia, a lice-ridden hut is home to an entire family. The cook stove belches smoke. Grains are mashed and made into injera, a type of flatbread with a spongy texture. A small, sad-looking garden provides some vegetables. Livestock roam about a little pen. The animals are a primary source of protein for this family. Otherwise, food comes from outdoor community markets. There, though, supplies are sporadic. The rains. The dust. They disrupt food chains. Transportation on rocky roads can't always be counted on.

Climate causes roadways and transportation networks to degrade more quickly. In northeast Africa, temperatures can range from freezing cold in the highlands to some of the hottest in the world in the deserts. Because there is an active volcano in the Afar region, the Earth's surface there is so hot it can melt the soles of your shoes. The heat also makes the land mostly infertile. The geological conditions are diverse, and determine, usually negatively, the fate of the poor country's citizens. Numerous NGOs from around the world try to help to make lives better.

Aid workers from World Vision, a global charity that operates a comprehensive network in the region, provide food aid where and when they can. Their mission is valiant. They tour desert regions where malnourishment is rampant. They check in on vulnerable families, like the one described. They educate visitors on the plights of the people who live here. But they can do only so much: 2.7 million people in the country are acutely food insecure. There are just too many doors to visit.

In a report, World Vision states that "the greatest food security challenges overall remain in sub-Saharan Africa." It notes that one in four people in the area lack adequate food.

The situation is tragic, but not lost. The factors that result in hunger are multitudinous. The World Food Programme (WFP), a branch of the United Nations that works in countries to deliver food assistance and improve nutrition, claims there are six major causes of hunger:

- **Poverty.** "People living in poverty can't afford nutritious food for themselves and their families. This makes them weaker, physically and mentally, so they are less able to earn the money that would help them escape poverty and hunger."
- **A lack of investment in agriculture.** "Too many developing countries lack the roads, warehouses and irrigation systems that would help them overcome hunger. Without this key infrastructure,

communities are left facing high transport costs, a lack of storage facilities and unreliable water supplies—all of which conspire to limit farmers' yields and families' access to food."

- **War and displacement.** "Conflicts consistently disrupt farming and food production. Fighting also forces millions of people to flee their homes, leading to hunger emergencies as the displaced find themselves without the means to feed themselves."

- **Unstable markets.** "Roller-coaster food prices make it difficult for the poorest people to get nutritious food consistently—which is exactly what they need to do. Families need access to adequate food all year round. Price spikes, on the other hand, may temporarily put food out of reach, which can have lasting consequences for small children."

- **Food wastage.** "One third of all food produced (1.3 billion tons) is never consumed. This food wastage represents a missed opportunity to improve global food security in a world where one in eight is hungry."

In the middle of this list are the effects of climate—the sixth cause of hunger. Climate change latches onto each and every issue on the list. The WFP calls it a "hunger risk multiplier." According to scientific papers published by the National Academy of Sciences, climate change will significantly increase the risk of food shocks and malnutrition worldwide. Lack of sufficient water for irrigation and higher ozone levels will share responsibility for the decline in supplies, along with other disruptions such as harsh weather. The UN's Food and Agricultural Organization (FAO), which monitors the world's food supply, forecasts food demands will increase 60 percent by 2050 as the global population reaches an estimated nine billion people. "Competition for land, water, and food could lead to greater poverty and hunger if not properly addressed now, with potentially severe environmental impacts," the FAO says.

This means the strides made since the new millennium are poised to be erased, relegated to the rearview mirror and the relatively better times of the past. The phenomenon is not exclusive to developing countries. In the United States, forty million Americans face hunger every day.

Activists are working the world over to wipe out hunger and make food more available. Warren Buffett's son, Howard, for one, is committed to stamping out hunger in America. He sees the challenges climate change will bring to the food industry. "It's an overwhelming issue," he told *The Atlantic* magazine in a revealing profile, in which he discussed how climate change has made weather more erratic, hurting the farming industry. His solution? Improve farming methods to increase crop yields. He is committing hundreds of millions of dollars to the cause. (Apparently he can afford the undertaking, with a reported trust and foundations granted from his father that are worth more than four billion dollars.)

Advances in soil engineering could, as we learned in chapter six, bring about new prospects, new ways of farming, such as those being promoted by Buffett. But there is no solution for the harsh equation of more people plus less arable land equaling a global food deficit. Retooled farmland, for all of its promise, won't be able to change that equation. So what if we reengineered food itself?

The first genetically modified organism was developed in 1973. The modification involved cutting a gene from one strain of bacteria and pasting it into another. A year later, scientists used similar technology to alter the gene sequence of a mouse. Both cases involved giving the organisms more resistant antibodies. Pharmaceutical companies then began experimenting with other forms of genetic engineering. But it wasn't until 1992 that the first genetically modified food was introduced into the market.

Flavr Savr tomatoes became the first type of GMO (genetically modified organism) approved for human consumption. The modification extended their shelf life and made them firmer. Other crops have since been bred to resist pests and diseases. Some of the most common GMOs are soy, corn, and sugar beets. GMOs also end up as feed for livestock, which of course then artificially conditions animals, although not their genetic structure.

Genetic interference is much more of a radical departure from nature than synthetic diets and growth hormone injections. Changing genetic codes means changing the building blocks of the organism itself and making it something else—something transgenic.

In 2009, a drug produced by the milk from a transgenic goat was approved by the US Food and Drug Administration (FDA). This was the first time a genetically engineered animal was introduced, albeit indirectly, into the human food chain.

All of this laid the groundwork and set the precedent for the first artificially engineered animal designed to be consumed by humans. Meet EO-1 Alpha. She is a salmon, but she was made in a laboratory at Memorial University in Newfoundland.

During the late 1980s, researchers there were studying antifreeze proteins in Arctic fish when they happened on another discovery: By inserting a growth hormone into the DNA construct of an Atlantic salmon, it grew twice as fast.

"You basically inject thousands of eggs. After the eggs are injected, you wait until the little organisms grow up. In this case, to be little fish. You look to see who took up the DNA and turned it into a protein," explains Ronald Stotish, the former CEO of AquaBounty, which patented EO-1 Alpha's DNA sequence and brought it for regulatory approval in the USA.

In general terms, EO-1 Alpha is known simply as the founder fish.

Because researchers were not particularly interested in growth experiments (they were trying to figure out how salmon withstand living in frigid waters), the accelerated size issue wasn't exciting to them. To entrepreneur Elliot Entis, it was. He had taken a meeting with the two lead scientists from the salmon project, who were pitching to him the idea of keeping farmed fish alive longer in freezing waters. At the end of the meeting, they showed Entis a photo of two of the fish they had genetically modified. One was bigger than the other. They explained what had happened to EO-1 Alpha: She had grown to five kilos (eleven pounds) in eight months. It takes a typical salmon thirty-six months or longer to reach that size. Entis figured he was onto something. He soon built a business around developing genetically modified salmon that require less feed, energy, transportation, and grow twice as big over the same time period as salmon born and bred in nature. Those economies mean prices at the counter are cheaper, as much as one-fifth or one-sixth the cost of the average retail price for Atlantic salmon.

The idea was and still is to create a fish-protein product for the masses. Two-thirds of the world's fish stocks are either at or near depletion. A synthetic fish could help shore things up, or so AquaBounty supposes. But as you might expect, there are major obstacles and opposition to serving genetically engineered food.

The first transgenic fish was actually developed in China in 1984. According to the Chinese Academy of Sciences' Institute of Hydrobiology, professor Zhu Zuoyan injected a gene that regulates human growth into the eggs of three thousand goldfish. Based on that research, a theoretical model and system was built for the genetic engineering of fish. Carp and other fish species have since been manipulated. But not for food. These have all been experimental laboratory tests. The fish are kept as research subjects, nothing more. This is why EO-1 Alpha is such a big deal. (She herself died in 1992. But her DNA lives on!)

Salmon in the wild spawn by swimming upriver. Males get to the spawning grounds first, where they fight for the best real estate: not too shallow, not too deep, just the right amount of gravel. Then the females arrive. They use their tails to dig small depressions called redds on the river bottom. Males joust for the opportunity to reproduce with a female. When they finally do pair up, they hover over the redd. The female lays her eggs while at the same time the male releases white milt that contains sperm. When the eggs and sperm mix, fertilization begins.

Of the thousands of eggs laid, only a few fish from the redd will survive to spawn another day. A female salmon may cover several redds over a few days with the same male partner. Males, on the other hand, can reproduce with several partners and, hence, seed lots of redds. But the same fate awaits both males and females if they are Pacific salmon: They die after spawning. Their remains wash downstream as food for bears and other animals. Atlantic salmon, on the other hand, can survive and return to the same spawning grounds for years.

The entire spawning process is a fascinating product of nature. AquaBounty's genetically engineered operation is very different: Eggs and milt are removed by hand and placed into a container for mixing. Breeding is selective, allowing only for females to be reproduced. This controls production quotas. The milt also comes from artificial engineering. Batches of selected young females are given testosterone. This allows them to milt later in life but renders their chromosomes strictly female. As a further safety net on the number of fish bred, the mix of eggs and milt is treated so the eventual embryo becomes incapable of reproduction. The embryo ends up in a tray and is sent to the manufacturing facility.

Albany, Indiana, is where trays of EO-1 Alpha's progeny will land. Here AquaBounty is set to produce tons of salmon a year for people to eat. The company couldn't have chosen a more unlikely place to breed fish. Albany is in the middle of farm country, more than eleven hundred miles away from where the nearest natural Atlantic salmon spawning grounds are, in Maine.

The forty-three-acre Albany site is where the fish hatch. Trays with the embryos are placed into what look like baker's racks and taken from trucks at the loading dock. One hundred thousand eggs resembling caviar will eventually be positioned over basins to hatch. Within forty-two days, the fish will hatch and begin to swim and will be taken to the nursery, a massive industrial space where rows of giant circular fish tanks are housed. Steel catwalks above connect to walkways where workers in lab coats and high rubber boots look down to monitor the processing equipment. The lighting is dim. There are vents, tubes, and hoses. Concrete floors. As many as twenty thousand fish will be put into a tank. They are divided and allotted by size to better portion feed. Four tons of feed will get distributed per day at the site. Biofilters remove ammonia from the refuse. Water is treated by ultraviolet light and ozone. To prevent contamination, fish are moved, or flushed, by tubes from tank to tank or area to area. Once they get to size, they are sent to the primary processing facility. There, an automated belt system stuns each fish in the head. Then their gills are cut off and they are bled out. The salmon are boxed, put on ice, and are trucked away. Twelve hundred tons of salmon a year can be processed this way by the Indiana site alone. That's a lot of fish. Think of it on a per pound basis: It's 2.4 million. AquaBounty also has facilities in Panama and Prince Edward Island.

At the time of this writing, only Canada was commercially selling AquaBounty's salmon.

In 2016, a law was passed requiring that genetically modified products be labeled as such. But the rules and language aren't uniform or clear, and are easily sidestepped. Disclosure information, for example, can be displayed as a bar code. You'd have to scan an item to learn if it is genetically modified.

Stotish, who took over as AquaBounty's CEO in 2008 when Entis stepped aside, recalled one particularly contentious public hearing where groups averse to genetic engineering showed up in droves: "People like the Center for Food Safety, Food and Water Watch, Oceana, Friends of the Earth, Earth Justice . . . got up and said, 'If you eat this fish, you will get cancer. If you eat this fish, your children will die. If this fish is approved, the world will come to an end. There will be an apocalyptic extinction of all salmon species all over the world.' And it just went on and on and on."

No significant health risks have been proven to be associated with AquaBounty's GM salmon. (Otherwise, assumably, the FDA and USDA would not have approved the fish for human consumption.) But as with all types of genetic engineering, there is alarm that gene splicing will give rise to mutant monsters.

About 40 percent of Americans believe genetically modified foods are more harmful to human health than other food types. Which, to be sure, is a big percentage. But that still leaves the majority of the US population believing, according to a Pew Research Center survey, that GMOs are the same or better for you than other food. (Forty-eight percent see no difference; 10 percent are pro-GMO.)

In Europe, GM foods must be clearly labeled. The EU views genetic modification rather unfavorably, and many refer to GM foods as "frankenfood." Japan, too, has rigorous standards for GM foods. And in China, as is the case with other countries around the world, standards are tightening. GMO labeling is on a path to be de rigueur.

Stotish believes that is unfair. He says wild fish can contain worms and toxins. "Should wild-caught fish be labeled 'this product may contain worms'?" he asks rhetorically.

"You probably saw the FDA approval of the ocular injection using CRISPR to reverse blindness in patients that have a congenital blindness. People are perfectly willing to accept that. When you talk about those sorts of applications, people are effusive and praise it and they think it's wonderful. If you talked about producing food that they eat a little more efficiently, they suddenly freak out. That's part of the education process. But if we're gonna meet those [food needs] in the future, we're gonna have to be able to communicate. We're gonna have to be able to have products that can meet those needs," he says.

CRISPR stands for Clustered Regularly Interspaced Short Palindromic Repeats, which is, in a practical sense, human gene editing. CRISPR technology can be programmed to target specific genetic codes in order to correct mutations or to treat genetic causes of disease.

CRISPR, too, has its critics. Ethicists say gene manipulation can go too far, and they have called for limits on the extent to which CRISPR technology can be used. Others claim that human gene editing bolsters the stigmas associated with different diseases and poses dire eugenic risks. But the flap over CRISPR pales in comparison to the heated opposition for GM foods.

Food & Water Watch, a nonprofit organization based in Washington, DC, says it champions healthy food and clean water for all by standing up to corporations that put profit before people. It is vociferous in its battle against AquaBounty, comparing GM salmon breeding to "the failed blueprints from *Jurassic Park*."

Another nonprofit based in DC, the Center for Food Safety, whose mission is to curb the use of harmful food-production technologies by promoting organic and sustainable agriculture, says AquaBounty's genetically modified salmon is a "dangerous experiment." And Friends of the Earth, a grassroots environmental group, believes the salmon will set off environmental impacts that will be impossible to turn back.

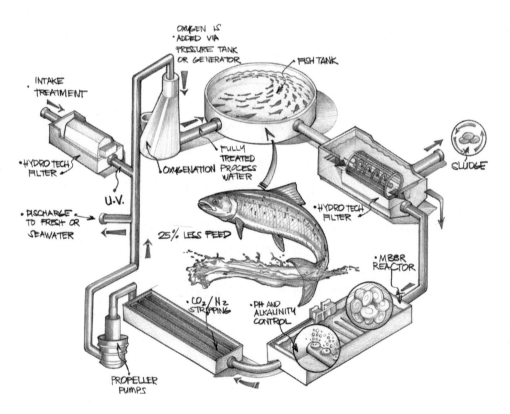

The AquaBounty Plant

Academic researchers largely disagree with the alacrity. Their biggest concern is that GM salmon will somehow be released into the wild, breed with other fish, and destroy entire marine ecosystems.

A laboratory experiment conducted in a controlled setting at Purdue University showed that when transgenic salmon are introduced to wild populations, the bigger GM salmon dominate and seize mates. That reduces offsprings' abilities to survive, and over generations could wipe out the entire species.

The chances of that happening with AquaBounty's salmon are slim. As Stotish notes, "we're on land." Moreover, all the salmon are female. And there are numerous safety checks. Even walking from one room to another at the Indiana facility requires signing in and out, boot scrubbing, and other protocols. Other facilities, however, may not be so strict or careful. And this certainly is a variable that can't be assessed.

The truth is that no matter the consequences, genetically engineered foods are likely to end up on our plates, or on the plates of people who are starving for food of any kind. The moral question of manipulating life-forms—playing God, if you will—will have to be weighed against a higher likelihood that more lives will be lost from starvation. It is not an easy answer. But as population growth continues, the solution may become clearer: genetically modified foods or famine.

Soylent Green

 Genetic modification is one way to create more food for the world's population—turbocharging natural processes. Another way is to make a food substitute.

"Soylent Green is people!" is the famous line from the 1973 science fiction movie *Soylent Green*, starring Charlton Heston. The plot revolves around an overpopulated world suffering from the effects of global warming (back then, commonly called the greenhouse effect). In the film, rations are provided, ostensibly giving people all their nutritional needs in pill form. The pill is called Soylent Green. Spoiler alert: the famous line stated above.

Today, Soylent is an actual food company. It manufactures "meals" in powder form. The powder contains all the protein and nutrients humans need to survive, the company says. The name is purposeful, snatched from the pages of the Harry Harrison novel *Make Room! Make Room!* on which the Charlton Heston film was based. Of course, today's Soylent, the real Soylent—available online and at your local 7-Eleven—isn't made from people. It's a highly thought-through mix of ingredients described as full, ready-to-drink meals.

"In a world with a rapidly growing population, and rapidly diminishing resources, we all need access to nutrition that is cost-effective and easy to consume. We're pro-GMO, pro-sustainability, and ready to change how the world thinks about food," Soylent, based in California, says.

The company refers to its food as "engineered nutrition." Every scoop of Soylent powder has twenty grams of protein, twenty-one grams of fat, and twenty-six different vitamins and minerals.

The idea for Soylent came in 2013 when software engineer and entrepreneur Rob Rhinehart was working at a start-up in Silicon Valley and didn't have the time, or the desire, to cook or shop for energy-enhancing meals. He began a thirty-day experiment, trying different mixes of ingredients to give him the sustenance and energy he needed in a single serving. Rhinehart came up with a formula that worked for him. A crowdfunding campaign followed. And now Soylent has premade drinks, coffees, and powders.

"Earth's population will reach 9.7 billion by 2050, and feeding that many people will require a 70% increase in food production. With 38% of land already used for agriculture and 41 million people in America alone struggling with food insecurity, finding solutions to our food access concerns cannot be ignored," the company says.

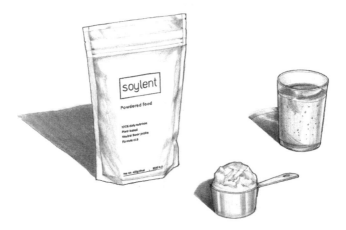

Soylent Food Powder

It's a world that eerily sounds like the one portrayed in the book and movie, but one (let's hope) that won't follow the same food plot.

Cultured Meat

Scientists have figured out a way to skip the "life" process and go straight to producing meat from animal cells.

"Lab meat" is cultured cells grown in petri dishes to form meat. Think of it as how algae grows into a mass of solid material.

Memphis Meats can make beef, chicken, or duck by extracting animal cells and cultivating them with sugar, amino acids, fats, and water in a laboratory setting. After about a month or so, the meat gels. Hamburgers, chicken nuggets, sausages, or other variations can be made from it.

The company refers to the process as "clean meat" because it painstakingly identifies the best animal cells to grow into a meat product. The clinical process also removes the need for antibiotics and growth hormones, which can be found in traditionally grown foods. The big jump in benefits, though, is going straight from lab dish to table plate and eliminating the entire process of raising and slaughtering animals for food.

"Cells are building blocks of all food we consume and at Memphis Meats they are the foundation of our approach. We make food by sourcing high-quality cells from animals and cultivating them into meat—think of a farm at a tiny scale. We cut some steps from the current process (like raising and processing animals) and bring nutritious, tasty meat to your table," the company advertises.

Traditional animal food processing is, of course, rife with environmental issues: consuming vast areas of land; necessitating huge amounts of feed stock; causing profligate waste and wastewater; and producing excessive greenhouse gas emissions.

Memphis Meats negates almost all of livestock's environmental issues. Its products will still require water, produce some waste, and use some electricity, so greenhouse gases won't be completely eliminated. But there are obvious benefits from growing food in a laboratory. Albeit, they come at a terrific economic cost.

The price for a pound of cultured meat is somewhere in the range of twenty-four hundred dollars. The goal is similar to that of electric vehicles: Volume will bring prices down. Demand is certainly there.

Meat consumption is poised to grow exponentially over the coming decades. The United States has reversed a decline in meat consumption and forecasts are for demand to increase in the next decade. Global demand is also on an uptick. Global meat consumption is expected to grow 73 percent by 2050, according to the Food and Agricultural Organization of the United Nations. Population growth and economic prosperity in the developing world are the primary drivers. As incomes rise in developing countries, meat—a more expensive protein—becomes a bigger part of household diets.

Synthetic lab meat could very well answer the question of "Where's the beef?"

CHAPTER 14

Toilet to Tap

Rivers feed the most fresh water to people on Earth. But their service isn't relished or respected. In fact, rivers are among the dirtiest and most polluted elements on the planet. They have become filled with toxic runoff, bacteria, and heavy metals that kill, mentally and physically impair, and breed birth defects.

For instance, the Citarum River in Indonesia, widely considered the most polluted on the planet, sees twenty thousand tons of waste and more than three hundred thousand tons of wastewater enter it per day. It's located just 80 miles south of Jakarta, the country's capital, and flows for 186 miles. Along its banks, two thousand textile factories drain their toxic runoff into it, raw and unfiltered. Meanwhile, twenty-five million people rely on the river as their main water source for drinking and irrigation.

The sacred Ganges River in India and Bangladesh is dying because of all the raw sewage that enters it—about 1.3 million gallons a day. The sludge is so bad that in certain stretches it causes the long river, which spans more than fifteen hundred miles, to run red. Yet, millions of people still dip and wash in the Ganges. For them, it is holy water.

The fifteen-mile-long Sarno River, in southern Italy, around the Bay of Naples, is so awash in pollution that it is considered a case study in geomorphology due to all of its untreated supply of effluent. The public health risk is demonstrable, but officials have failed to keep the water out of municipal supplies. That means residents and tourists alike are exposed to its hazards.

While not surpassing the magnitude and legacy of all those water travesties, the 344-mile-long Flint River in Michigan may have captured the most global attention for its toxicity and infection of drinking supplies, exposing thousands of people to dangerous levels of lead.

The Flint water crisis, which lasted from 2015 through 2017, has become the subject of countless news stories, documentaries, and television programs. Numerous campaigns have also been produced to help those affected. Celebrities and pop culture figures such as Will Smith, Cher, Eminem, and Madonna, among a slew of others, have donated money and sent hundreds of thousands of bottles of water to help people. Activists and influencers such as these highlight the fact that it is the poor and disenfranchised who were put most at risk.

The Flint story is one mostly of money and politics. To save costs, the city government in 2014 decided to change its water source from Lake Huron and the Detroit River to the Flint River, which runs through town. But proper water management was sacrificed. Dangerous levels of lead were discovered in the Flint River supply, exposing one hundred thousand residents, including as many as twelve thousand children, to the toxicity.

Standing over the Flint River on a walk bridge, the muddy and brown cyanobacteria-ridden water flows slowly underneath. A couple of ducks paddle around and three buoys bob just before a cascade. On the tranche below, where the river widens some before it bends and slips out of sight, a few oil patches not more than a foot long float. The gray and blue slicks dance around each other. They break off in different directions and cast their own shadows on what lies beneath. You can't see bottom. The river is thick like pea soup. An empty plastic water bottle is washed ashore, discarded under an open drain. A sewer is on the opposite bank of the river, sixty steps across.

In less than ten minutes you can walk the water line from the Flint River, up Harrison Street, and to the municipal buildings and water district offices. The city itself isn't going to win any modern design awards, and many of the office buildings and shops are empty or abandoned. It looks as though they have been that way for a while. But there is a real community feeling to the downtown area. The numerous houses of worship from myriad faiths—Episcopal, Presbyterian, even a Masonic Temple—showcase active event schedules. People apparently come together frequently to talk about beliefs.

Charlotte (she offered only her first name), a security guard standing post at one of the old buildings under construction downtown, says the water situation is getting better. When the lead toxicity was discovered in 2015, many residents were left without safe supplies. Bottled water was the only option.

"Most people have water now," Charlotte says. She grew up and still lives in Flint. What concerns her are the effects of the harmful water on the children. "It's the kids. We adults, we can get by. But the kids, with the lead and whatnot . . . that's what's so bad."

Children are particularly susceptible to lead poisoning because their young and growing bodies absorb four to five times more of the toxin than an adult would from a tainted source. Lead poisoning affects the brain, liver, and kidneys. It is stored in teeth and bones, which are also harmed as lead accumulates over time. Poorer children who are undernourished are harmed the most because they typically lack strength-building nutrients, such as calcium and iron; they end up absorbing more lead instead.

Charlotte points to some of the renovations and new construction taking place around town: the new outdoor café where the old movie theater used to be, a strip of retail outlets, the farmers' market. The development and redevelopment are hope for a new future. Flint residents, of course, are hyperaware of the water problem they face and the worldwide attention it has received. Charlotte even discusses the dramatic film *Flint*, starring Queen Latifah and shown on the Lifetime channel, that was based on the water crisis. "Oh yeah," she says. "I've seen them all."

It's good, the attention. It shines a light on problems that should not exist. It wrings out the tainted water issues of which so many of us are unaware. What's in our water?

Outside the restroom area on the University of Michigan campus in Flint is a small notice. It reads, "Is the water at the University of Michigan-Flint Safe?" The notice explains how the university filters and tests its water, and how it is helping the local community deal with the crisis.

In Flint, there is seemingly easy access to water. After all, the namesake river runs through it. When the municipal water, expensive to pipe in from the Detroit agency managing it, was threatened with a shutdown, who could blame officials for looking at a source closer to home? It doesn't excuse the mismanagement or poor practice standards. But it is a telling example of what the future may bring if water supplies run thin or run out.

Since no one has been able to figure a way to manufacture water, the only option is to use the supply that we already have—even if it's tainted, even if it's toxic, even if it takes walking miles to retrieve.

Planet Earth's water supply hasn't changed much since the ice age. What has changed is how much we use and squander. Between the nineteenth and twentieth centuries, when the world's population quadrupled, water use spiked by a factor of fifteen.

With close to eight billion people on the planet and climbing, water demands are stressed and will be strained further over the next fifty years on population count alone. There will be nearly ten billion human beings on the planet by 2050. That means the amount of water needed for everyone on Earth to survive and thrive will increase exponentially. And that's just on the basis of our personal consumption. The industries we create also gulp water supplies. Nearly half of all the surface fresh water in the United States, for example, is used to generate electricity. Most of the rest of the fresh water is for our food and agriculture. However, technology is beginning to consume significant amounts of water, too. In the US alone, data centers, which allow us to access and search the World Wide Web, use more than six hundred billion liters (or about 1.6 billion gallons) of water per year. That's like putting ten water bottles in the hands of every person on the planet. As technology proliferates, so will water use.

Besides all the water that is used, we also are polluting more and more of it. A United Nations world water assessment program found that the majority of sewage in developing countries is "discharged untreated, polluting rivers, lakes and coastal areas." And pollution isn't just a problem in the developing world. All around the globe, water pollution is on the rise.

There is an estimated 332.5 million cubic miles worth of fresh water on the entire planet for the taking. It sounds like a lot, but as one giant drop, it would all fit well within the borders of the United States. Think about that size as compared to the rest of the globe; it's puny.

Most fresh water is found in ice caps and glaciers. Another 30 percent is in the ground. Less than one half of one percent resides in surface waters—lakes, rivers, streams, swamps, and the like—where it can be readily accessed.

Climate change is further crimping that supply by allowing more moisture to float around in the atmosphere rather than falling to Earth as precipitation. As global temperatures rise, warmer air becomes less dense, keeping more water afloat, metaphorically speaking anyway. The science is complex. But the point is that rising temperatures will cause quicker evaporation, leaving ground sources with less water.

There is, by nearly all scientific accounts, plenty of fresh water for everyone on the planet to use. That said, by 2050 it's forecast that nearly every population center on Earth will experience some kind of water scarcity problem. The issue is access—accessing it to drink, bathe, and for sanitation. As many as two billion persons—or about one-quarter of the world's population—are affected by availability. It's one thing to have a sufficient global water supply. It's another thing to ensure it's potable and can get into the mouths of people who need it. This was perhaps most efficiently explained in chapter 12. But there is good reason to discuss tainted water supplies in more detail, too.

Arid areas are expanding. The United Nations projects that by 2030, half of the world's people will be living in desert-like lands. Without water, people will likely move closer to readier sources, and that will result in millions of "water refugees." A high estimate puts the total number of people displaced due to a lack of safe water at seven hundred million.

As it stands, dry lands comprise more than 40 percent of the land surface on the planet. There's not much room for more.

There are famously dry places in the world: sub-Saharan Africa, China's plateaus, Australia's outback, and the southwestern United States, home of the legendary Dust Bowl. Sure, people could, as the comedian Sam Kinison used to yell, "Just move!" But this magnitude of people moving would not be such a simple task. Other solutions must be explored.

In the 1960s, Orange County, California, realized it had to do something about its dwindling water supply, or it would run out in twenty years. Orange County is between Los Angeles and San Diego and experiences a semi-arid climate, forcing it to rely on groundwater and the trickling Santa Ana River that runs through its borders. In California, water distribution runs from north to south, with communities downstream left with less and less supply as their northern neighbors take their shares first.

After World War II, Orange County transitioned from an orange-cultivating community, as its name implies, to a housing and business center because of the oil and aerospace industries that had grown nearby. So what to do to allow for population expansion—a place for all those employees to live? Find more water.

Engineers began exploring desalination because of the coastal community's proximity to the Pacific Ocean. Desalination is the process of removing salt from water. And because the oceans are so vast, housing so much salt water, some see the seas as the future source of fresh water.

Drinking salty water, such as seawater, alone can be deadly for humans. Kidneys need to process water that is less salty in order to hydrate the body. Too much salt, and dehydration occurs. Eventually, too much salt water ingestion will result in death.

There are several ways to remove salt from water. The oldest method is heating salt water to produce steam or condensation that can be captured as fresh water. Sailors have used this method for centuries. It requires a lot of energy and isn't particularly efficient, as might be understood, because it involves chasing drops of humidity.

Another more modern method of desalination is ionic, using electricity to separate out the salt ions and produce potable water. Again, this is super energy intensive.

Reverse osmosis is yet a third method. Here, water is forced through a membrane that filters out the salt. It is effective, but largely useless for seawater because it can't deal with high salt content. There is just too much salt to filter out.

Orange County engineers were highly attuned to using vast amounts of energy. Pricing for public water supplies is sensitive. The energy for desalination methods is costly, and as a result, engineers began looking for other solutions. They began investigating wastewater recycling.

Wastewater recycling involves a whole other filtration process than desalination. The process is commonly referred to as "toilet to tap" and involves treating sewage water so it becomes clean enough to drink. The yuck factor is often the biggest issue to overcome with this method. It's a public perception and psychological challenge, not so much a practical obstacle.

Wastewater recycling requires a large number of steps of filtering and treating and testing. So much so that the end result is often fresh water that is "cleaner" than other sources.

Most wastewater is washed from drains and sewer pipes out to sea. That means there is a massive opportunity to be had for cleanup and to convert all that waste into usable, potable water. The Orange County Water District realized this opportunity early on in the 1950s and '60s and eventually decided to team up with its next-door office neighbor, the sanitation district, to see if it could manufacture a sustainable source of water for residents.

"It was a coincidence of infrastructure," says Shawn Dewane, a resident of the area and a director of the water district, who recites a bit of history at the water department's offices. He says the office locations' happenstance gave officials more opportunity to discuss recycling and treatment ideas.

Through a program dubbed Water Factory 21 (the "21" standing for twenty-first century), an advanced wastewater treatment process was launched in the mid-1970s.

Today, Orange County is home to the largest water reuse treatment plant in the world, servicing 2.5 million people. It has nearly closed the loop, re-treating most of what is flushed down toilets and reconditioning it as potable water. People, or rather their waste, have indeed become the solution to Orange County's water problem.

Recycled Wastewater

Water industry professionals, public utility officials, and government leaders from all over the world come to visit the sanitation and water departments in Orange County to marvel at how they operate and pump out fresh water from waste.

A tour of the water treatment facility includes seeing (and smelling) the raw sewage pipes that carry waste from sewage drains into giant rake filters at speeds of up to five miles per hour. Rags, condoms, plastics, and other large items (even a once-rumored bowling ball) are removed and placed on a conveyor belt. After that, the sludge moves through a grit chamber where eggshells and coffee grounds and the like are removed with a high-pressure air system. Next, foul air is captured and funneled into a large silo. There it is mixed with soda and bleach to get rid of the smell. After this, the sewage water is ready for primary treatment.

Huge basins slow the water down to a trickle, allowing solids time to float to the surface, where scraper arms skim the top and bottom of the basin to get rid of most of the materials left. The water is then sent to more trickling filters and aeration basins to further clean it. Next it is sprayed over honeycomb-type materials where bacteria grow. When the water trickles through, microorganisms eat any solid materials that weren't filtered out. Finally, the water is sent from the sewage treatment plant to the water district plant next door for more processing.

When the sanitized sewage water first enters the water department facility through its massive pipes—one hundred million gallons flow through them per day—it goes through a microfiltration process using bundles of polypropylene fibers to remove contaminants. The next step is reverse osmosis, where the pressurized water is forced through sheets of tiny membranes. After it's purified, the water is exposed to high intensity ultraviolet light to disinfect and destroy organic compounds. Because it has been cleaned so thoroughly, minerals have to be mixed back in. At this stage, the water is potable. But it isn't distributed for direct use. Instead, water is funneled to groundwater basins, where it gets mixed with other sources and is naturally sifted over approximately a six-month period before it makes its way back into water pipes for people to drink, bathe, or flush their toilets with; the cycle continues.

———————

The sink is the last stop on a trail tracing raw sewage through the various filtration and treatment processes. It's an industrial, stainless-steel sink with a small faucet—something that might be seen at a restaurant. The water that pours out from the tap looks clean and fresh. The small, plastic tester cup is clear and so is what's in it: the recycled wastewater. There is no scent to it. However, there is still a slight hesitation before taking a sip. The images of raw sewage running through the pipes are too fresh, perhaps. Yet down the hatch the H_2O goes. It tastes fine. In fact, there is no taste to it at all.

Of the 2.5 million residents who drink from the same proverbial well, there have been no formal complaints about the water. There have been no instances of curious outbreaks or diseases. But that doesn't mean there haven't been and aren't contamination issues with which to be dealt.

A form of acetone, a compound commonly used for cleaning liquids but which is relatively harmless, nearly snuck through Orange County's water treatment system. Mike Wehner, the assistant general manager of water quality and technology for the water district, says the acetone made it past the mechanical treatment system but was filtered out by nature's own environmental buffer (that six-month percolation process of rising up through the local aquifer's sand and silt) before making it into groundwater supplies.

Another filtering issue is for NDMA, the chemical N-Nitrosodimethylamine, something they are continuing to work on. NDMA, which is a carcinogenic, apparently has the ability to form again after it is purified. It is commonly produced as a by-product of industrial chemical processes. Wehner says the county is managing post-treatment conditions to minimize re-formation of NDMA. And a water department spokesperson says an advanced oxidation process of ultraviolet light with hydrogen peroxide took care of the problem. Still, concerns abound.

Water testing is a highly regulated business that leaves no room for mistakes. Miss something and people can die. Flint, Michigan, is a case in point. The change in water systems left a gap in testing.

All around the world, people are exposed to poisons due to improper water treatment.

In Bangladesh, more than a quarter of the population—forty million people—have been poisoned by arsenic in the public water supply. In southern France, millions of people drink water polluted by pesticides. They and the growing number of other instances give pause to trusting municipal supplies.

Studies reveal that water pollution is out of sight and therefore mostly out of mind. The *Washington Post* on June 8, 2016, published a commentary on water pollution in light of the Flint water crisis that said, "Citizens tend to focus on problems they can see and experience. . . . This means [they] are less likely to notice water pollution in routine encounters—only foul smells and strange colors in the water lead to complaints."

Orange County long ago discerned this public condition and it's the reason why the water district goes beyond testing and screening to render safe water supplies. Public education is critical, it realized—especially when it comes to wastewater.

Jim Herberg, who runs the Orange County Sanitation District, says educating people and businesses on what they should and shouldn't flush down the toilet or dump into sewer drains makes a huge difference in prescreening toxins. His agency manages a "What 2 Flush" media campaign. "It's simple, the toilet is only meant to flush the three P's— pee, poop and paper." The agency informs people through handouts, community events, public outreach talks, and school programs, among other ways to spread the word.

Next door at the water department, officials want to expand their use of wastewater to service the entire county's population of 3.2 million.

"We are very much committed to recycling as much water as we can," says Wehner. He brings up the fact that stormwater, treated wastewater, and the like are still in large amounts discharged into the ocean. They're "thrown away," he says, by other municipal providers. "We didn't want to do that. We thought, 'Let's make use of it.'" And they have. And they want to continue.

Recycled water is an imperfect way to describe wastewater reuse. All tap water is actually already recycled naturally, as water molecules break apart and are put back together in concert with the planet's hydrological cycle. Water from the middle of oceans or high atop mountains is eventually lofted into the atmosphere, spread about, and forced back down to Earth as precipitation. That's how mountain snow can eventually become ocean rain: through nature's long process of reconditioning two hydrogen atoms and one oxygen atom bound together as H_2O.

Toilet-to-tap schemes, however, largely take nature and its hydrocycle out of the equation. The natural water cycle can sift out toxins through sand and soil before delivering it cleanly as potable. Only time will tell whether we humans can do a better job of breeding out impurities. Meanwhile, wastewater recycling is catching on in major cities as part of a comprehensive menu of water options for communities. The United Kingdom, Australia, Belgium, Singapore, South Africa, Israel, and Namibia are some of the nations experimenting with wastewater reuse.

William Sarni, a renowned water thought leader and consultant who advises major corporations and municipalities on their water use, says he likes the prospects for wastewater because it's cost-effective and predictable. He says wastewater's biggest obstacles remains public perception.

Sarni has written a number of books on water. His book *Water Tech: A Guide to Investment, Innovation and Business Opportunities in the Water Sector*, outlines myriad systems and ways for people and businesses to be better water stewards. He writes and speaks of the urgent attention water is in need of around the globe. But no matter the urgency, in some places wastewater treatment and reuse just won't stick. It has nothing to do with the cleanliness of water. It has to do with perceptions of cleanliness itself.

In drought-ridden Australia, the opposition group Citizens Against Drinking Sewage (CADS) serially protests wastewater reuse projects. The name of the group defines its mission. Members just don't think that it's right for humans to consume wastewater. Yet, with limited ways to reuse and recycle the world's water supply, wastewater is something that we all probably will have to get used to and, no doubt, drink during our lifetimes.

In the future, the water we use in our homes may be recycled over and over and over again in a closed loop system where drains and pipes filter, treat, and heat wastewater, redirecting it back to taps and shower heads and hoses. At the same time, our solid waste can be diverted and used as energy (biofuel). The Orange County Sanitation District does this and saves millions of dollars per year in energy costs.

Reduce, reuse, and recycle may not be practices aimed solely at hard products anymore. We'll have the same opportunities with waste and water. The "gross" factor is something we'll just have to get over. There simply won't be enough basic resources such as water to go around for everyone in the future without rethinking how we can tap our toilets for more than flushing.

Water from Air

 We all know solar panels can convert sunlight into energy, but a company has found a way to convert sunlight into water.

Zero Mass Water's SOURCE technology needs only sunlight and air to make drinking water. Four feet by eight feet in size, hydropanels look similar to solar panels and use advanced water-capture technology to produce as much as ten liters of water per day, or the equivalent of twenty regular-sized water bottles.

A standard SOURCE set is made of two panels and a storage reservoir that holds sixty liters (about sixteen gallons) of water. The system is completely off the grid, utilizing the power of the sun and a small battery to operate. The storage reservoir can connect to your plumbing and deliver water directly to your tap. It can also act as a water-delivery system in even the most arid deserts, making SOURCE technology a game-changer for communities that lack access to fresh water.

The World Health Organization reports that more than two billion people lack safe drinking water at home, and nearly a billion people don't have access to fresh water at all.

SOURCE applies thermodynamics, materials science, and controls technology to effect water capture. Essentially, the hydropanels capture heat and humidity that exist in the air and wring water from it.

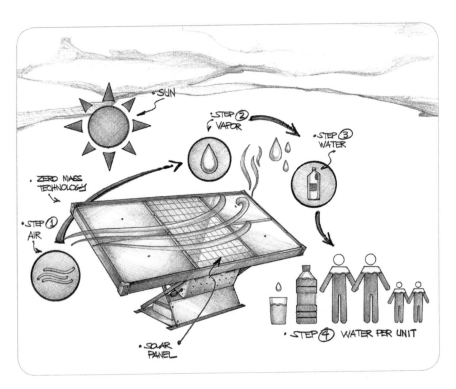

A Zero Mass Water Panel

Zero Mass Water, based in the United States, says its mission is to make drinking water an unlimited resource. It's also attuned to taste and quality. "SOURCE water is mineralized with calcium and magnesium, achieving optimal taste. With this perfect combination of electrolytes for your hydration, SOURCE brings the highest-quality water for your health right to your tap," the company says.

SOURCE's hydropanels have been installed all over the world, from developed, urbanized environments to rural communities. The system costs several thousand dollars, but pays for itself over time. Or as Zero Mass Water calculates, "With a fifteen-year life, your SOURCE will provide you with [on average] a case of delicious water per day, or 12 standard bottles, for less than $1.00!" The company also has a program, Water for Life (W4L), that enables customers (or donors) to help families or communities in need acquire a SOURCE system.

No longer is it up to nature and the weather to bring water from the sky. Zero Mass Water has hacked the atmosphere, skipping precipitation, and enabling water delivery directly to us on the ground.

Fog Catchers

 Fog is a cloud close to the ground, a mass of water particles that drift and morph or linger near shorelines or even in desert regions come morning. Temperatures yawning to rise and dew can make air thick with moisture.

Some people who live in dry places are familiar with fog catchers—fine nets that trap moisture as water droplets pass through them. Caught in the net, the droplets trickle down and are stored in tanks. In the Atacama Desert in Chile, the world's driest known desert, fog nets are used as a water source for entire communities. People there even brew beer with the extra water they capture. The US military uses fog nets for troops stationed in arid zones. These are just a few examples.

Depending on the weather conditions and the site, fog nets can capture an abundance of water, several thousand liters a day in some instances. Fog carries between half a gram to one-tenth that amount per cubic meter, depending on cloud density. Hoist one big net or many big nets and little drops turn into full tanks. Still, the nets are not very efficient. Lots of water gets lost in the capturing process.

Engineers at Virginia Polytechnic Institute and State University, better known as Virginia Tech, have improved upon the fog net design and come up with a technique that can capture three times as much water as traditional fog nets. Called Fog Harps, these water harvesters utilize only vertical wires, as opposed to the vertical and horizontal mesh patterns found on fog nets.

Virginia Tech scientists discovered that the horizontal aspect of the mesh disrupted the droplets' flow downward. By eliminating the horizontal material, water drops flow more freely. The size of the mesh patterns, they found, is also important to maximize water capture.

"Under controlled laboratory conditions, the fog-harvesting rates for fog harps with three different wire diameters were compared to conventional meshes of equivalent dimensions. As expected for the mesh structures, the mid-sized wires exhibited the largest fog collection rate, with a drop-off in performance for the fine or coarse meshes," they reported. Too big a gap and not enough water was caught; too little a diameter, and the water droplets were blocked.

The Fog Harp obviously stems from the shape of the fog catcher—all vertical wires like a harp. In drought-stricken areas, Fog Harps may begin to strike a better chord.

HUMANS V.
THE FUTURE

CHAPTER 15

The Artificial Intelligence of Self-Driving Cars

 This is the worst traffic in the world. A comprehensive study that examined traffic patterns of more than a thousand cities around the globe named Los Angeles as the most congested.

That study found that drivers spent 102 hours sitting in traffic here per year versus 91 hours in both Moscow and New York, the next most congested cities on the planet. In terms of countries, traffic in the US ranked worst overall in the developed world.

The worst section of the worst city's traffic is on Interstate 10 eastbound, between Interstate 405 and Interstate 110. It's where the west side of Los Angeles begins to break east; the beach cities of Venice and Santa Monica left behind for communities of the concrete jungle: Inglewood, Culver City, and Koreatown, on the edge of downtown L.A. Industry and commerce are here, less so residential neighborhoods. A peek over the guardrail reveals strip malls and strip clubs. Parking lots and industrial lofts. On a Friday evening, cars crawl by these sights. Google Maps says it will take 115 minutes to go 18.3 miles. The roadways on the GPS are highlighted in red, meaning slow going or generally congested.

The Rosa Parks Freeway, the local name for this stretch of highway, has four lanes that are dead stopped. The High Occupancy Vehicle (HOV) lane, reserved for those with two or more people in a single vehicle, or those with special permits and which is supposed to make traffic flow more easily, isn't moving either.

Workers leave large companies located on the west side of the city and head home to the east side of the city, where housing prices are cheaper. They end up at the freeway crossing at about the same time each weekday. It's what makes things so backed up.

During the 1984 Olympics, which L.A. hosted, city officials coordinated with large employers in the area to stagger work hours in order to lighten traffic loads, which in turn reduced smog. Athletes could breathe better. Fewer cars idling means less concentrated areas of pollution. Smog, as we've learned, can be produced from tailpipe emissions. But there is no Olympic-like traffic coordination these days. Businesses are left to their own scheduling. And that means five o'clock marks the end of working days and the beginning of mad rushes to get home.

Pollution is at its worst during nighttime rush hour because air gets warmed all day, and by the time late afternoon comes around, toxins are exacerbated by the heated conditions.

On a Friday evening in winter at 5:15, there are no vehicle accidents, but the 10 is clogged. There is a sea of red brake lights. Cars jockey for position, hoping to get ahead and move faster. It's a giant tease. The lanes where vehicles are inching along end up halting to a stop after a few feet. Nobody is getting in front of anyone else in any meaningful way. The only movement on the roadways is by motorcycles; they are allowed in California to travel between lanes.

A kid in a car seat in the back of an SUV stares up at a plane flying overhead. The back window of a limousine rolls down. Schoolchildren press their noses against the windows of their bus. There is time to notice such things at full stop on the freeway.

Movement eventually comes in waves: six miles per hour, nine miles per hour, twelve miles per hour—and then back down the scale. Seventeen miles per hour feels like speeding.

The sun begins to set and the orange glow is caught by the windows on high-rises in downtown Los Angeles. More buildings are under construction. They reach high, some seventy or so stories. L.A. is erecting a new future for itself, centered around the downtown area. It may change the decentralized nature of the city, whose famous freeways were built as progressive solutions to suburban living during the boon times of the 1950s, and beckoned car buying. California still has more car dealerships than any other state. The historic and massive investment in public roads over decades coupled with automobile and oil industry interests, have made it difficult politically for L.A. to wean itself off cars. Besides, people, polls have found, just downright prefer to drive from one place to another rather than use public transit. (Less than 10 percent of residents use the L.A. Metro system.) But a tipping point may have been reached. In 2016, voters approved a new tax to fund expansion of rapid bus, bike, and rail systems. Easing transportation burdens, however, is far off. It will take forty years to build all the infrastructure for new and improved public transportation options. In the meantime, if migration trends prove true, L.A. traffic is going to be worse. By 2040 a million more people are expected to reside in the city.

All over the world, traffic is getting bad, despite better maps and driver-assisted technologies. In 2012, drivers in Los Angeles spent less than sixty hours in traffic. And L.A. was the most congested city then, too. By 2050, the average driver in the average city will spend more time in traffic than a Los Angelean does today.

More people living in cities isn't helping delays. Until this decade, as we now know, most of the world's population lived outside of urban areas. Employment and opportunity have coaxed millions into densely populated areas.

Traffic dents progress. Lost hours of labor equate to more than three hundred billion dollars, or about fifteen hundred dollars per driver, in the US alone. Healthcare expenses zoom, too: Traffic congestion adds to air pollution that equates to higher rates of illness and medical costs.

Transportation is the second-biggest cause of global warming. If we can change the ways in which we travel to lessen our emissions, we might be able to create a back-door climate fix.

Roadways may be the worst engineered blight humans have made on the planet. And a lot of the ugliness can be blamed on the Romans. In Ancient Rome, around the year 312 B.C., the first major highway was built—Via Appia, or Appian Way. It connected Rome to southern Italy and was used to transport military supplies. Over time, other long, straight highways were constructed to expedite the movement of the Roman army. As military conquests continued, so too did Roman roadwork, extending throughout Europe and Northern Africa. It's estimated that the ancient Romans built fifty thousand miles of highway, and that's the reason for the saying, "All roads lead to Rome."

Some of those roads are still used today. But if you've ever tried to drive in or out of central Rome itself, you'll see how poorly the roads have kept up with traffic—it's madness. Ancient Rome may have pioneered city planning and roadways, but modern Rome, in modern times, has been labeled the most dangerous city in Europe for traffic accidents.

To be sure, Ancient Romans can't be blamed for not envisioning automobiles and population growth. But today's governments and urban planners can be blamed. They chronically fail to account for population increases and driving patterns. In Los Angeles, for example, a multibillion-dollar project to expand its major freeway, the 405, won't curb congestion, construction officials admit. It will merely stay it, at best. The extra lanes being made will only coax more drivers to the roads. What's at work is something a new breed of transportation researcher calls induced demand. It means when newer, wider roads are built, more people flock to them and drive on them, believing things will move along faster. But the added drivers actually increase traffic. Yet, this analysis is rarely taken into account by urban planners.

Such examples of stunted vision are why a group of students in a Stanford University laboratory dreamed up a world where artificial intelligence takes over the management of our transportation needs. How we get around, or move from point A to point B may not be up to us soon; algorithms will decide.

Drive.ai is the brainchild they came up with to change the future of transportation. The "ai" in the company's name highlights its mission to harness artificial intelligence to transform transportation as we know it. And it's all about autonomous, or self-driving, vehicles.

Autonomous vehicles are theoretically safer, more efficient, and help reduce traffic congestion. Even just a single self-driving vehicle amid a sea of human drivers can significantly reduce traffic. A University of Illinois study shows when a traffic jam begins by, say, a human driver slowing down too much or braking too hard, the autonomous vehicle controls its speed more intelligently. That means the vehicles behind it don't slow so much or brake too hard. Flow is managed better. And there are fuel savings. When a vehicle slows and speeds up again, it consumes more fuel than if it maintains a steady speed. The study revealed that fuel savings amounted to forty percent in the test group. Imagine not just one AI vehicle, but a fleet of them navigating more expertly through streets and highways and facilitating traffic flows. Congestion would be lightened, environmental impacts lessened, and let's admit it, tempers tamed.

What makes self-driving vehicles so smart is the information they rely on to make decisions. It turns out that when it comes to driving, computers are a lot smarter than humans.

Dustin Hoffman, playing the mentally impaired character Raymond Babbitt in the film *Rain Man*, repeated over and over the line, "I'm an excellent driver." The same can be said of a lot of people: They think they are good drivers, but they really aren't. Example: texting. Multitasking while driving has become part of the human condition. Accidents have skyrocketed due to texting alone, never mind talking on mobile phones. A driving analytics company found that drivers spend 3.5 minutes on their phones either texting or talking per hour of driving. Driving under the influence, exhaustion, and other human failings also support the theory that computers are better suited to take control of the wheel than we are. Which is why robots are set to take over the roadways. In 2013, back in the lab at Stanford, the founders of Drive.ai glimpsed this vision of the future.

"We were working on all sorts of different robotics projects," recalls Drive.ai's cofounder, Sameep Tandon, who is still filled with youthful exuberance about the technology. "This seemed the most obvious." He smiles frequently and welcomes questions. Dressed San Francisco casual in an open-collared shirt and jeans, he still seems amazed at how quickly his Mountain View, California, offices are growing. (In just a few years, Drive.ai staffed to more than one hundred employees and took over an entire office floor.) In 2019, the company was acquired by Apple.

Drive.ai is considered a leader in the autonomous driving space. It has hatched a deal with Lyft for an autonomous ride-sharing program, and it was one of the first start-ups in California to receive an autonomous vehicle testing permit.

While at Stanford, Tandon and his lab mates knew they had the chops to develop a commercially viable AI product. The question was what to focus on. After a lot of research, they zeroed in on self-driving cars. The reason was, and continues to be, the potential that autonomous vehicles bring to different aspects of life; unexpected potential, too.

If self-driving vehicles catch on as believed, parking lots will be a thing of the past. So, too, will be garages. Certain aspects of delivery services and post office functions will be dinosaured. (Order online and send a car to pick it up.) And efficiencies will breed a 10 percent reduction per vehicle in carbon emissions.

Replacing recreational drivers, your average private citizen, is not really what autonomous driving is all about. The biggest play is in the commercial sector, with ride-share programs, technician routes (such as cable installers), and delivery services. These involve fleets of vehicles operating on expected routes. This makes programming easier. The easier the programming, the more efficient vehicles can become.

On a very basic level, artificial intelligence is software. Once the software is coded, it can begin to teach itself. For example, if other vehicles continually flow into a left lane at a certain time each day, and the right lane proves faster despite the appearance of stopped traffic, the autonomous vehicle will eventually just stay in the right lane.

"When you think of the self-driving car, the first thing is the factors on the vehicle. These cars come equipped with cameras, LIDARs, radars, ultrasonics. What these sensors provide is like a three-hundred-and-sixty-degree understanding of what's happening," Tandon explains. A LIDAR (light detection and ranging) system works likes a radar but utilizes light instead of radio waves for detection and surveying.

"In our vehicle, we have about ten cameras. There's four LIDARs and a radar system. This is the basis of where we begin. After that, there's a lot of software capability. The first thing on the software side is what we call perception. You're just trying to have the car understand where things are around it. For example, where are other vehicles? Where are bicyclers? Where are pedestrians? By having these vehicles understand what's in their environment, we can then have them make accurate decisions, as well as put them on the road so they can drive safely. The next part is the motion and decision-making.

"Now, the trickiness in all of this is that if you're trying to build an engine, an AI engine for the car, traditionally what you'd have to do is write rules for what the car should do in different scenarios. If it's raining, you should expect the vehicle to have to deal with the reflection on the road, which messes with perception of stuff. If you're at a stop sign and you're trying to figure out who has right of way, typically it's hard to write all these things out as rules. What we realized a couple of years ago was that an AI machine learning approach was going to be the best way to actually get all this knowledge into our vehicle. We're telling the vehicle, so annotating the data set, what decisions it should do in various situations, and then training the AI system so that given brand-new situations, given brand-new data, it's able to really generalize to the more challenging problem," Tandon explains.

For example, in traditional rules-based software programming, an image would be scanned into the program and a directive written—an image of a red light and "stop," to put it simply. AI has the vehicle do the learning itself instead of just abiding by the written rule. Hundreds of different images of stop lights are captured (in fog, snow, rain, etc.) and myriad scenarios play out for the software to learn from. What to do when a light is flashing red, for instance. In addition, there isn't just one vehicle out acquiring data, there are hundreds. That allows the program to learn rapidly, exponentially. Mapping systems and guidance systems like Waze and Google Maps work in a similar manner. The more users the better, and the better the system can inform drivers about traffic patterns and guide individual drivers more efficiently. It's crowd-based learning.

Combine a guidance system with autonomous driving and far-out strategies can increase traffic flows. "Go right, then left, speed up, and then slow down" are all instructions handled without the human disruption factor—say, rubbernecking an accident.

The tagalong effect of self-driving vehicles is even more disruptive. People can live farther outside of urban areas and work on the way to and from the office without the hassle of actually driving. This could flip the phenomenon of more people migrating to cities for opportunities, which is the cause of so much congestion. It's a big idea.

"I think there are a lot of places where self-driving car technology will end up revolutionizing things," Tandon says. "If you think about it, there's no need for things like drive-thrus anymore. Food should come to you on demand. The UPS model where you have a package delivery happening every day, that doesn't really make sense, either. Again you should have packages come to you predictably as you need them or whatever." He says storefronts may begin to be designed differently to better allow for drop-offs and pickups. The *Wall Street Journal* reported that Zara, the clothier, has already found major success with its system of utilizing retail locations as convenient pickup and return points for online customers.

Residential real estate in suburbs may be designed without parking garages. Public transportation schedules could be coordinated with ride shares and personal vehicles. Efficiency will prevail.

Still, there is the fear factor. Most people are afraid of new technology, especially technology that takes control of anything away from them. Add to that the physical dangers associated with driving, and many people balk at the prospect of autonomous vehicles, although that may be changing.

Hesitation about handing over driving control to a computer doesn't occur when you first get in and buckle up. It's when a patch of pedestrians crosses the driveway that you are exiting. "Does it see them?" "Do they see us and understand that this is a driverless vehicle?" "Will we stop?" The anxiety continues at red lights, left turns into approaching traffic, lane changes, and when vehicles cut in front and do something unexpected, like brake suddenly. Soon, though, after some comforting experiences that exhibit the vehicle does indeed know what to do—how to drive—the anxiety fades and the fact that a computer powered by AI is making life-and-death decisions becomes no big deal.

Still, there is evidence that computers won't have perfect driving records. Self-driving vehicles have been involved in numerous crashes.

Although Drive.ai itself has not experienced any accidents on public roads, Tandon admits that accidents are inevitable. "We'll never be perfect and to hold the industry to a zero-death kind of standard," he admits. But he also believes autonomous vehicles will learn to be safer. "We start with the human standard and then over time we think that we'll be able to surpass the human standard and build safer systems."

When compared with the number of traffic accidents and the loss of lives with human drivers at the wheel—approximately 1.3 million deaths per year and as many as 50 million people injured—computers have a lot of room to do a better job.

Drive.ai's test car is an electric-blue Nissan NV200 tricked out with rooftop cameras, sensors, LIDAR, radar, and other high-tech devices. It is actually a retrofitted New York City taxi. Streaks of yellow paint left on the inside of doors, window frames, and runners give its ownership history away. Sitting in the backseat and looking at the monitor affixed to the seating panel gives some clues as to what the Drive.ai software is processing: grainy, red heat sensor–looking images of people (pedestrians we are passing by), traffic lights, major structures such as bushes and trees, obstacles on the road, barriers and lanes. Another screen with CCTV-like images shows in video form what's going on in front and on the sides of the vehicle.

It seems as though all the possible views and angles are covered. But Drive.ai also has a backup to its AI system, called Telechoice. It's human operators who, back at headquarters, track vehicles' movements and can override the system's decision-making in tricky situations. For instance, if one of the autonomous vehicles in the fleet comes upon a construction site

A Drive.ai Vehicle

or other barrier on the road, it will likely stop—and stay there. Proprietary directions to drive around or make an unusual driving maneuver will probably not have been coded into the vehicle's knowledge base. Therefore an operator can take over and guide the vehicle around the obstruction.

Eventually, Drive.ai executives foresee tapping into community and government transportation systems and working with first responders to inform guidance. This could, if also adopted by other self-driving systems, alter the "drivescape" of the planet. Roadways have not been built to be directionally efficient. They have been built around centers of commerce and population that may be outdated. Guidance systems such as Waze and Google Maps have proven this true. Once-traffic-free residential zones have turned out, in some places, to become more convenient pathways for commuters. Of course, this makes residents in these more efficient commuter zones batshit crazy because of the traffic influx. But as computers take over routes, emotional consideration is lost.

There are twenty-one million miles of roads on Earth. It is a maze of constructed paths that interrupt, interfere with, trample on, and rip up the natural landscape. Artificial intelligence will in all likelihood remake the world and, in doing so, reroute our roads and our traffic patterns. There will be a cost to all of this, of course. Disruption always brings about damage to the status quo. That means transportation jobs will be lost. It means commuting schedules may change and work hours altered upon efficiency recommendations. And there is a psychological effect: We'll give up control to a part of our lives.

This last bit has Wendell Wallach, author of *A Dangerous Master: How to Keep Technology from Slipping Beyond our Control* and a scholar at Yale University's Interdisciplinary Center for Bioethics, very concerned. "It is taking autonomy away from humans," he says. Sparkling with intelligent quips and professorial explications, Wallach zeroes in on what he believes will be the biggest obstacle to the transportation efficiencies brought about by self-driving vehicles: political stability. Job losses, as witnessed throughout history, breed malcontents. If driving technology results in mass unemployment, as it might—the trade, transportation, and utilities sector is the largest employer of people in the United States—then society is in for an abrupt confrontation.

There are already groups vehemently opposed to self-driving vehicles. In New York, the Upstate Transportation Association, an organization that represents professional drivers, including taxi and livery drivers, is lobbying for a fifty-year ban on self-driving vehicles in the state. It says autonomous vehicles will cost thousands of jobs and harm the state's economy.

"There are a lot of people with guns in their closets who may choose to go out and sit on highways and say 'bring it on,'" Wallach cautions. (He says this as a general point, and not specific to the New York group.) While he believes artificial intelligence can no doubt make our transportation needs and demands more efficient and safer, he doesn't know whether we are quite ready for handing over control just yet. He says, "The artificial intelligence industry wants us to believe that its widespread use is inevitable." But he himself doesn't believe in inevitability. "We really don't know how we are going to move forward," he says.

An AI winter could set in whereby technological advancements will slow whilst adaptation and comfortability among people waits. Or, he equivocates, things could speed up. But before "we get too far down the road," as he puts it, and artificial intelligence controls more and more aspects of our lives, he would like to see an ethical commission put into place that sets rules governing technology's place in our lives. He is working on assembling such a global oversight organization.

Drive.ai's Tandon is fully aware of the fear factor and the artificial intelligence education gap that exists among the public. First, he doesn't believe there will be a massive loss in jobs due to technology. "Other jobs we never even thought of will come about," he predicts. But he does agree that more public education is needed. To that end, Drive.ai holds demonstrations and test drives to get people comfortable with its products and "letting go" of driving responsibilities. "We are at the very, very beginning of all this. It's like when the Internet was first formed. Things are going to take off in so many directions," Tandon believes.

Allowing artificial intelligence to take over our roadworks and transportation systems could ultimately transition the entire human design of the planet. If getting from point A to point B changes, will that change the way point A and B look? Will the means recast the ends? The Silk Road, once the dominant trade route in Asia, has willowed to a dusty reminder of history and inexpediency. Many of the cities along its former bustling route have faded into nonexistence.

Cities of the future are taking AI and transportation into consideration. They are looking at a total urban redesign, a city where self-driving vehicles are complemented with public transportation options. Places where walkable streets and promenades exhibit better qualities of life. Fewer parking lots. More green space. Cleaner air. Those that aren't moving toward intelligent designs, artificially intelligent designs, are likely headed for a crash. Self-driving vehicles are moving ahead, fast.

Defying Gravity

We can get to Boston from New York City in twenty-six minutes, Dubai to Abu Dhabi in twelve minutes, and Los Angeles to Las Vegas in half an hour. The Hyperloop allows us to do this: zip along at speeds exceeding six hundred miles per hour. Some versions will have us go even faster.

Hyperloop is gravity-defying technology that propels pods at breakneck speed through tubes. Pods can hold cargo or people. The technology is based on the sealed tube creating a vacuum, and therefore a frictionless environment for transportation. Because the capsules float, they can speed through tubes, much as how old pneumatic chutes sent mail within a building. (Although Hyperloop technology is vastly different.)

The celebrated entrepreneur Elon Musk popularized the idea of a Hyperloop, and now there are several versions. "Hyperloop consists of a low pressure tube with capsules that are transported at both low and high speeds throughout the length of the tube. The capsules are supported on a cushion of air, featuring pressurized air and aerodynamic lift. The capsules are accelerated via a magnetic linear accelerator affixed at various stations on the low pressure tube with rotors contained in each capsule. Passengers may enter and exit Hyperloop at stations located either at the ends of the tube, or branches along the tube length," Musk explains. His original executive summary, design, and concept paper are all available to the public. In fact, Hyperloop's technology is open source, meaning anyone can have a go at it, or riff off it.

The Hyperloop idea reportedly came about from Musk's frustration at transportation options from Los Angeles to San Francisco. They all suck for various reasons, ranging from trip time to expense to convenience. Hyperloop is meant to fill a void and provide wicked-quick transportation inexpensively. Another bonus: It's climate friendly. By using solar panels, Hyperloop is self-powering.

What Hyperloop does do to the Earth, however, is dig it up. Tunnels or pylons have to be dug between end points to lay the tube.

The Hyperloop provides an ideal transportation option given the cumbersome logistics of flying by plane; the relatively slow technology of trains; and the traffic congestion problems with vehicles.

Hyperloop technology has its obstacles, too, of course. Fastening oneself inside a subsonic tube and barreling along at hundreds of miles per hour is one. Acquiring land is another. Funding is yet another. And safety and security add to the long list of challenges. Still, Hyperloop reshapes how goods and us people can get from one place to another, faster than ever, on Planet Earth. It provides a fifth mode of transportation for humans, adding to vehicles, planes, boats, and trains.

As Musk says, "Short of figuring out real teleportation, which would of course be awesome (someone please do this), the only option for super fast travel is to build a tube over or under the ground that contains a special environment."

Perhaps next, teleportation?

CHAPTER 16

The City of the Future

Every degree of global temperature rise incites consequences of calamitous effect. A 2.7°F (1.5°C) increase, and the planet will experience chronic heat waves of Saharan intensity. Double that, and the Amazon rainforest will collapse, glaciers will melt more rapidly, and forests will begin to grow in the Arctic. Without carbon emission abatement, temperatures could theoretically soar even more. That would be the doomsday scenario, where cities become thick with smog and pollution; disease runs rampant; fresh water is difficult to source because of droughts; food becomes scarce; traffic congestion becomes nightmarish; marine life dies off; extreme weather becomes chronic, with more intense storms; floods push communities inland; and resource-lacking vicinities vanish as livable options.

As it stands, the Intergovernmental Panel on Climate Change reports that unless carbon emissions are slashed by 45 percent by 2030—a nearly impossible feat—the 2.7°F/1.5°C rise is assured. That means we must recast how we are going to live in the future. We will be forced to adapt to a new, more hostile natural environment and create livable cities. For urban planners, architects, and developers, this compels a redesign of cityscapes and services.

In Abu Dhabi, the capital of the United Arab Emirates, such a redesign is already under way. Here, famed architect Lord Norman Foster has laid out plans for a prototype for the future city of the world. It's called Masdar City, and it's expected to be completed by 2025, with fifty thousand people and fifteen hundred businesses taking up residency in the 2.5-square-mile development. To date, two thousand people have moved in and call it home. These are future dwellers—climate colonists, let's call them. They live in an extreme environment on the fingertips of the harsh Arabian Desert, and yet they thrive. Air, water, energy, and food have all been manipulated, engineered, or reengineered to make life more comfortable. Buildings are strategically positioned in harmony with the sun to keep them temperate. Smart windows capture energy and funnel it to storage cells in basements. Air ducts redirect wind and cool rooms. There are no power switches. Skylights provide natural light. Motion detectors turn on LED lights when needed. Water pours from recycled systems. There are agricultural lands and test labs of vertical farms to produce local food products. The entire city is connected to the Internet. Autonomous electric shuttles get people around. Green parks abound.

Masdar is the only city in the world that has been specifically developed to mitigate the harm we've inflicted upon nature and to completely leverage the power of the sun. The hope behind it is for others to be inspired, copy its engineering and designs, or put their climate-killer products to the test.

Masdar is actually the name of a company. It is a subsidiary of the Abu Dhabi government's Mubadala Investment Company. Masdar was begun in 2006 to advance clean energy innovation in Abu Dhabi and around the world. It is an important link for the UAE, a petroleum state, to diversify beyond its vast reserves of oil, which one day will run out, if "peak oil" observers are to be believed, or if the world weans itself off fossil fuels.

Masdar City was begun as a physical test center and a green-print for sustainability. The mission has been and continues to be demonstrating the commercial viability of renewables.

To make the city come to life, Foster and his revered architectural

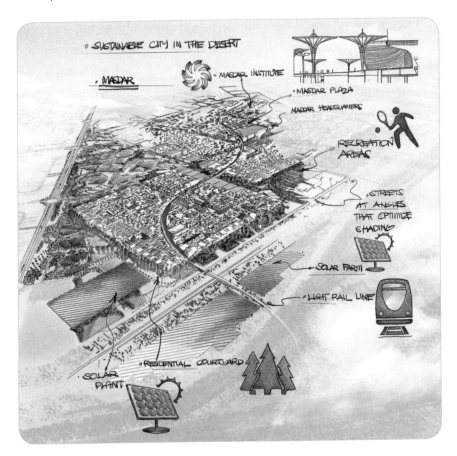

Masdar City

firm, London-based Foster + Partners, was tapped. Foster has designed megaprofile projects all over the world, and is one of the most esteemed architects of our time. A winner of the Pritzker Architecture Prize, the top honor in his field, he worked with Steve Jobs to create Apple's campus in Cupertino, California, among numerous other high-profile and admired projects that blend high-tech sensibility with sustainability. One of his more eye-catching buildings is the Swiss Re building in London, commonly known as "The Gherkin" (because it looks like a cucumber).

Elegant and still internationally active in his eighties, Foster answers questions from his London base, after having spent the summer in the United States. He says the Masdar project is special because "it has allowed us to realize a community that brings together many of the issues relating to architecture and sustainability that have been driving us since the late 1960s. We felt privileged to be working with a visionary client to push boundaries, question assumptions, and think about new ways that we might live in a future without fossil fuels and powered only by solar." Masdar still has a ways to go before it's complete, or 100 percent solar powered, for that matter. The city is an odd mix of the bold, the new, perhaps the future, with an aesthetic nod to the past.

A wicked sandstorm has just passed. Visibility dims and objects far away can barely be seen. There is dust everywhere—on roads, sidewalks, buildings, and equipment. Masdar grows out of this amalgam of dirt, stone, and steel. It's an oasis, really, on the outskirts of Abu Dhabi—a sustainable community surrounded by cities built for other types of grandeur and on the backs of profits from oil production.

There is a small sign indicating that it's there, Masdar, down a small road next to the airport. But there is little else to hint at where this city of the future might be. Coming upon it on a bright summer's morning, what you see first are the residences: six stories tall, white with colorful panels irregularly placed on them. The windows on each unit are tall and narrow. These are the housing units for students attending the Masdar Institute of Science and Technology. They, too, play a role in the future; they are encouraged to not only study but create and innovate in the field of sustainability.

Construction equipment is everywhere. Crews in hard hats and orange vests appear in pits. Building plans are unfolded out in the open, pages flapping in the air as foremen look at site plans to confirm what goes where. Despite the 111°F temperature, the city is bustling with workers, students, executives, and foreign delegates looking to co-opt ideas on sustainability that they can adopt back home.

Parking lots are full. They are covered to shade vehicles from the sun. There is a sparkle to the light-colored walkway material that takes you up a short stack of stairs from the lot to a shaded entrance area. Floors rise above you. Buildings are built on pilings. The space beneath allows for cool, open air to circulate. You can walk from structure to structure this way and be unaffected by the throbbing heat of the sun. The heat here doesn't so much hit you, as it might stepping outside on a hot day in Florida. Middle East heat swallows you whole and chokes you. Avoiding it is a constant battle.

Masdar isn't really a city. It is a community, more akin to a college campus than a commercial hub. Chris Wan, who heads the design teams at Masdar, says the "city" name was misappropriated because of bad translation. The same word in Arabic can be used for both city and community. In fact, there does seem to be a strong sense of community here, where people are working toward a similar goal, the end game of which is a model for future living.

Not all of Masdar's advancements, of course, will be able to be adopted in other places. This neck of the planet is unique. Foster says that there is no one-size-fits-all solution to the design of cities, and culture and geography are what make each city special. For instance, Masdar's trade, culture, and customs are not remotely the same as, say, New York City's or London's. But what can be adopted are more general sustainable practices, plans, and products.

Bosch, the German conglomerate that produces everything from electronic equipment to kitchen appliances, is piloting its Climotion solution inside buildings here. The technology uses artificial intelligence to optimize air mixing. This makes air conditioning and ventilation much more efficient and effective. It results in energy savings of up to 30 percent and, of course, that translates into less carbon dioxide emissions. Or take the windows. They are made from electrochromic glass, also known as smart glass, which can change from clear to opaque depending on the intensity of the sun. Building materials themselves are even considered for their eco footprint. Zinc is used on roofs because it takes less energy to manufacture and is easier to recycle. Timber is used for structure because it is a renewable resource. The shapes of buildings are also important. The library is built in the shape of a baseball cap so it is shaded from the sun on its sides but allows light and air to flow under its bill.

One of the most immediate realizations about designing a city from scratch is that it has to have a shape to it first. The shape of buildings can then follow. Masdar is being built as two square shapes, one large and one small. Northwest–southeast is the dominant orientation of the streets. Sun, shade, wind, and walkability are all taken into account to make the most out of the patches.

There is a red tinge to the entire place. Local sand is mixed into building materials and it looks like burnt clay. The streets are made of porous material and feature a beautiful mosaic design that allows drainage in the barely distinguishable gaps. More stunning are the bright rust-colored buildings that rise above. The buildings are just a few stories tall and have solar panels on their roofs. But they are works of art. Balconies curve in tandem, open only at their far sides. They are walled but have large cutouts at their far ends. There is an intricate, repeating matrix pattern through which air can flow. These are the faculties of ancient Persia wind towers. They allow for both breeze and shade. They capture air naturally and whisk it inside.

The buildings themselves are sited close to one another. This produces shade in the streets or alleyways between them. Without air conditioning, old desert cities leveraged smart design. Human intelligence overcame harsh natural habitats.

Foster spent months in the region researching ancient cities and how they battled the elements. He took his inspiration from the architects of the past to build something advanced, far ahead of its time. Masdar is built on the premise of adaptation and resiliency. It's a place for climate survival.

"Climate change is a phenomenon that affects humanity as a whole, and we all share a responsibility for the environment. Masdar is, to our knowledge, the only experiment in the world that explores how a community can be sustained twenty-four seven by solar power—and in this case an energy-intensive scientific research facility," Foster says.

Nature, of course, has historically aimed its threats at us, particularly in certain places: the deserts, mountains, jungles, and angry reaches of the ocean. Foster wisely took his inspiration from communities that have long figured how to live among wicked elements and survive the worst weather and climates thrown at them.

"Our starting point was to look historically—before the age of cheap energy where you have power at the flick of a switch—at how local communities tamed the extremely harsh climate of the desert. Learning from the history of indigenous, vernacular traditions, we can see that they tended to be quite compact to create naturally shaded streets and buildings. They were orientated in such a way that they captured breezes, they were lifted slightly above the desert floor and they would grow upwards. When you look at the detail within those communities, you can see that there is a combination of shade, outdoor spaces and greenery, with pools of water that allow evaporative cooling through the settlement. As you penetrate the interiors, you find the larger houses connected to wind towers that capture the cooler, higher breezes and they pull that cooler air into the dwellings. Beautiful detailing of the interiors also captures and frames views out while creating a high degree of privacy—a very beautiful tradition.

"We applied these learnings to Masdar where, like historic desert settlements, it is scaled to be very protective, offering shade, water, and a degree of intimacy—all the qualities that we perhaps associate more with the villages and the towns on a tourist trail. We married this with cutting-edge innovative technologies such as the highly insulating ETFE [ethylene tetrafluoroethylene] cushions used in the laboratory walls, or the ten-megawatt solar field that powers the entire settlement through renewable, green energy. The buildings therefore combine innovative technologies with traditional methods to evolve a new built form that responds to the harsh desert climate, while accommodating contemporary functions," he says glowingly.

The United Nations has taken to investigating ancient indigenous knowledge, too. The UN Office for Disaster Risk Reduction (UNDRR) is studying indigenous cultures to find out how they build, how they adapt, and how they survive in extreme conditions—over centuries. Certain communities in the Philippines, for example, look to natural signals as early warning systems for typhoons: the shape of waves, the smell of the ocean, cloud colors, how plants and animals react. They also build with resiliency in mind—on higher ground and with resistant materials. Prevailing winds are accounted for as are high-tide lines and flood zones.

Here in Abu Dhabi, they pay close attention to the sun and the sand, the wind and where fresh water lies. Building structures for the future means not only accounting for the history of the climate but also the future of the weather.

Wan says his team queues off Foster's detailed designs. They take a holistic approach to engineering in order to address practical realities. "High-performance buildings have to care about aesthetics, as well as comfort as well as the resources used," he says. It's an integrated design process from the beginning.

What he learned over the decade he's been there is that smart homes, innovation, and technology all take a backseat to one thing: planning. "The real experiment is not what you see, it's the process," Wan says.

Urban design is a collaborative process and the biggest hurdle is educating all the key principals from different disciplines and keeping everyone on the same page. There are many disciplines involved, including construction engineering, landscape design, interior design, lighting, acoustics, and more.

Masdar is built around three pillars: environmental, social, and economic. All the projects have to make sense financially, or else they are not implemented. This produces building and design models that are replicable and aren't produced merely as academic exercises.

Wan says there has always been between a 5 and 15 percent premium for building green. "The thinking is green buildings have to cost more," he says. But the construction budgets for Masdar are the same as typical designs. "It's how we do it that is challenging industry norms," he says. That is why all that advanced planning he spoke of is key.

Masdar is a people-centric city, which is important because it emphasizes walking. But the climate has to be accounted for. The distance between the city center and residential buildings, by example, is calculated by how far you can go before you begin to sweat.

Drawing from the past, buildings have thick walls (for insulation) and small windows (to block the searing heat of the sun). But innovative planning brings these tactics into the modern era. Between 30 percent and 40 percent is the ratio of a building's openings to solid mass, Wan says. Anything less keeps things too dark inside. Anything more, and it allows in too much heat. He says the engineering art is optimizing traditional knowledge with modern know-how. This relates to the materials chosen, too. Sure tin and timber, as previously discussed, have their attributes. But are they produced locally? The transportation of goods has to be factored in as well for carbon emissions and carbon footprints. Take concrete. Masdar uses recycled concrete as much as it can. But local supplies are low. So they won't fly it in.

Water use is managed with low-flow features, and Masdar is experimenting with water-storage possibilities. For example, it is capturing condensation from air conditioning as a reuse source. It already engages in gray water recycling and wastewater recycling. (Gray water is water used by sinks, showers, tubs, and washing machines. Black water is sewage.)

Wan notes that those are all easy steps—commonly used by environmental practitioners throughout the world. Masdar is looking at technologies that will better store and restore water—or even do without. Other steps include recycling construction waste and, of course, squeezing energy savings wherever possible. That means reducing demand and utilizing renewable supplies. Wan says this is why buildings are begun as passive designs first and technologies are add-ons.

To conceptualize how savings could be had across waste, water, and energy—the three prongs of environmental sustainability—Wan and his team designed traditional structures, and then doggedly asked each other questions about how they could make them better. The questions, Wan says, were the tough part. The design is relatively easy. Preplanning this way created a blueprint where architects took what they could make spatially and then made it work for their specific desert environment. This strategy produces designs that are real, not conceptual.

That also jibes with Foster's vision. "From the outset, the settlement was also conceived as a testbed for the technologies being developed in Masdar's own laboratories—our masterplan is flexible to accommodate these advances," he says, adding that the next phases of the development will draw on lessons from the Masdar Institute campus. "We also learn from our own projects."

This fluid design model allows more sensical elements to be incorporated as lessons are learned over time. The underground transit system is a good example. It turns out that the original technology has been surpassed, so plans have changed. There is now an aboveground autonomous shuttle for "people mobility." Or take some of the solar energy issues. Rather than a single solar field transmitting energy in bulk, engineers found that certain solar panels were more effective placed on top of buildings. This also produces a local benefit. The excess energy produced by individual homes can be carried forward as an energy credit for the residence under a government utility scheme. This is a big benefit for homeowners and gives more incentives to conserve energy.

Food, too, is a big experiment. Wan says they are testing sustainable farming, which is a tall order in the drought-stricken desert. Vertical farming also comes into play, but with a twist: Masdar is looking at ways to use vertical farm rows as green walls and shading devices. "We try to look at things as doing more than one job," Wan says.

In one to two years, Masdar plans to have its food security plans well-mapped out.

Foster says it is important to design cities that are flexible and can adapt over time to change. "The design of Masdar is deliberately non-prescriptive and looks to adopt changing technology in the future. It illustrates an approach—of working with nature, learning from traditions, and capitalizing on the advancements of clean technology to create a sustainable urban model for the future," he says.

But will it work?

Wan walks around a giant table upon which is a full-scale model of Masdar. There are miniature buildings, roadways, parks, and public areas. We are a floor below the building's lobby, but there are floor-to-ceiling windows and doors to the outside and the subterranean plaza. The temperature is kept cool by the lack of direct light. The original autonomous vehicle—the one phased out—is parked outside by the door.

The office color, a washed slate, and decor are bland. If it weren't for the self-driving vehicle, we could be in pretty much any office park anywhere. But this is in the desert, in an exotic part of the world, where Western sensibilities have only recently been embraced.

Wan is an animated speaker. Thin and bespectacled, he is passionate about the project. He moved here from Hong Kong, and his understanding of space and design hails from the challenge of congestion. Hong Kong is one of the most crowded cities on the planet, with more than seven million people crammed onto an area of just 414 square miles.

That works out to seven thousand people packed into every half mile of land area.

Another well-known architect and designer, William McDonough, is also from Hong Kong. He says growing up there shaped his environmental outlook. McDonough created the Cradle to Cradle concept, which proselytizes the idea of building recyclable and sustainable materials from the outset so that every product can be reused or made into something else again and again (cradle to cradle, as opposed to cradle to grave).

Wan says when McDonough visited Masdar, they bonded over their Hong Kong experiences. With scant outside resources and such a huge population's demand for things, it should be no surprise that the mantra "reduce, reuse, and recycle" was widespread in Hong Kong.

City living, dense and reliant on resources outside the urban boundaries, is—like it or not—the environment of our future. This will be how the supermajority of us live—on islands of innovation, tricking nature to meet our needs. Whether climate innovations work or not will be up to time, and how much or how little we choose to act on climate change. As the saying goes, it literally will be a matter of degree.

To design a city built for the future from scratch is the chance of a lifetime. Foster knows this and it is why he is so engaged in the Masdar project and its outcome. He gets to map the future, according to his own vision for it. He's both grateful and hopeful. And he is also reflective. He puts into words decades of experiences and synthesizes a conclusion that is as elegant as his life's works: "My recent public talks have concluded with some visions of cities of the future. They show what New York, London, and San Francisco might look like fifty years from now. They are green, pleasant, flexible—they are centered around people not cars or infrastructure.

"I often make the point that these images are full of elements from the science fiction of my youth. This is not wishful thinking or fantasy. By way of example, I showed images from my past, of a local library, cinema, camera, typewriter, telephone, and post box. Could I have ever imagined the contents and power of all those resources and buildings shrunken into one device that I could hold in the palm of my hand—a smartphone?"

The things Foster refers to are, however, luxuries. Reimagining our future from today is being forced upon us by necessity. Time spans for adaptation to climate change are being condensed. We are facing new realities. Cities, our most concentrated forms of collective living, can—must—lead the way to resiliency.

We'll never go back to living in a wild and natural world. Our urban collectives—concrete jungles—where the most number of people live on the planet, ensure that. But that doesn't rule out bold new ways of living led by invention and innovation that put the power of nature in our hands.

Foster in his youth might never have imagined a smartphone. Can we imagine what a smart world might be in the future? Hopefully this book has given us a peek into what's possible and how we can correct our course toward a new climate beginning.

NEOM: A Post-Oil-Era City

NEOM is Saudi Arabia's attempt to pivot out of relying on petroleum to secure its place in the world. NEOM is a half-trillion-dollar plan to build a city that showcases the best that technology and innovation have to offer. It's a daring attempt to not only design the most advanced urban environment in the world but to position Saudi Arabia as a leading force for what's to come. It defines itself as "the world's most ambitious project: an entire new land, purpose-built for a new way of living."

Saudi Arabia says NEOM will operate as a country within a country, with its own tax and labor laws. It says the model city, more than ten thousand square miles in size, is a new blueprint for sustainable life on a scale never seen before, where inventiveness shapes a new, inspiring era for human civilization. Robots are expected to play an active role in the city's construction and ongoing service arena, from home deliveries to caring for the sick and needy. Artificial intelligence, 3-D printing, virtual reality, smart devices, and the Internet of Things (IoT) will all be embraced.

To be sure, NEOM will look cool, with artist renditions showing gleaming glass buildings and skyscrapers set against a backdrop of smaller white structures that arch around green spaces, bridges, and waterways.

NEOM plans to pioneer the future in sixteen sectors: energy; water; food; biotech; manufacturing; technology; mobility; sports; tourism; entertainment, culture, and fashion; media; design and construction; services; health and well-being; education; and livability. What will evolve is a post-modern development on the northwest corner of Saudi Arabia in the Tabuk region near the border of Jordan that faces Egypt across the Red Sea.

If the city is erected on schedule, it will be a future made quick: NEOM's planned opening is for 2025.

Conclusion

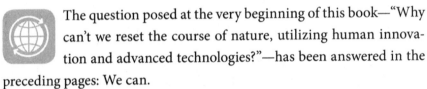The question posed at the very beginning of this book—"Why can't we reset the course of nature, utilizing human innovation and advanced technologies?"—has been answered in the preceding pages: We can.

That answered, next is forging the will to move ahead with those innovations, experiments, and technologies that allow hope to creep back into the conversation about climate change. For too long climate change messaging has been pessimistic, revolving around frustration, desperation, exasperation, and world end. The "we're all going to die unless we do something" salvo has fallen flat. Global carbon emissions continue to rise. Words are not being translated into necessary action. This, after decades of scientific warnings, reports, books, news stories, museum exhibits, political campaigns, international climate change protocols, and United Nations conferences. All of what we have read, seen, and heard come with the same ending: pivot to more eco-friendly practices, or perish. Yet that ending rings hollow.

Paralleling conventional messaging are the same old mitigation and adaptation measures, too: embrace renewable energy, lessen fossil fuel use, live more simply and more efficiently. Nothing refreshing beyond fiery discourse shouted by political alarmists is being promoted.

There is still wide-eyed belief that once people understand the errors in their environmental ways they will make amends and change—become more environmentally friendly. The hope is that sacrifice will reign over selfishness and self-interest. That's a bad bet. We eat unhealthy foods, drink beverages that aren't good for us, smoke cancer-causing products, and chronically ignore medical safety guidelines. You can't get much more personal than your own body. If people choose to ignore advice even about that, how likely is it they will heed environmental warnings about things that are largely unseen, such as carbon dioxide, and may happen at some time in the future, such as major sea level rise?

Geoengineering and the technologies that can immediately effect change are being sidestepped. Fear, I've come to believe, is the reason why. People are petrified of the unfamiliar.

When Edward Jenner in 1796 developed the first human vaccine (for smallpox) many people refused to take it long after it was proven safe and effective. They feared injury or death. The same holds true for geoengineering and methods devised to control nature and keep it in check since we've corrupted it so much.

There is also skepticism about putting the fate of climate change into the hands of the business and financial communities—those willing to take investment risks on new technologies. But that is what needs to happen.

This book is a clarion call for inventors and investors—society's underwriters of what's possible. Governments, our historic benefactors, are paralyzed by bureaucracy. The private sector has even taken over charting new frontiers in outer space, once the bastion of nation states. The mechanisms that can craft a more hospitable climate, which we can all enjoy, are languishing on the sidelines as wonkish debates slog on about how to most effectively create intergovernmental environmental policies to reduce carbon emissions.

If you are on the frontlines of climate change, then you are maddeningly frustrated. Extreme weather is getting more extreme and taking more lives. There are more risks to public health. Climate refugees are increasingly making camp. Our environs are becoming just as much threatened by inertia and conventional wisdom as they are by environmental pillagers.

The technologies laid out in this book are our best hope for addressing and reversing climate change. We should, by all means, examine their risks, but we should not hamstring their rewards.

The United Nations Environment Assembly is looking to further examine and set rules for geoengineering experiments to thwart quick adoption of climate intervention technologies. Some believe this is a pretense for quashing climate engineering progress altogether. This brings to the fore the main arguments in favor of and against geoengineering. Those opposed to geoengineering say it undermines efforts to cut carbon emissions: If "silver bullet" solutions abound, prevention efforts will stall. Those for geoengineering say prevention and mitigation efforts—small green steps undertaken by everyone—have either failed or won't remedy the climate quickly enough.

The environmental pendulum has swung from the climate deniers on the far right opposed to carbon emission mitigation to the alarmists on the far left opposed to far-reaching "quick fix" solutions. A mutual offense is needed to defend ourselves against the hostile environment we've created. That means grassroots activists should continue to encourage people to take small, everyday steps to help save the environment and policymakers should continue to enact rules that encourage carbon mitigation and adaptation measures. At the same time, the science and business communities need to come closer together and bring to market sober environmental products and programs that can redirect the crash course we are on with nature. The quicker we admit the Earth—our common Mother—is sick and can no longer care for herself, the sooner we can intervene and help her heal. Natural remedies aren't working.

Medical vaccines have saved humans from dying off in droves. Geo-engineering and environmental technologies are the vaccines I believe we need to cure the planet's ills. It's time to take our shots.

Acknowledgments

There are so many people to thank for the words in this book. But there is one person above all others who should be thanked most. Before a book becomes a book, it's an idea. And sometimes that idea isn't fully flushed out. Sure, you can have a strong concept for your thesis. Sometimes, though, that concept can lose its way before it becomes a story. My longtime literary agent Susan Raihofer slapped that story out of me—pushing back on my writing and the book's direction until clarity reigned. I can't thank her enough. Usually, when a book is done we'd celebrate with a glass or two of wine at lunch at The Blue Water Grill in New York City's Union Square. Alas, that restaurant has shuttered. Instead, good fortune prevailed and we happened to be in London at the same time to knock back a pint together at The Punchbowl pub. Now the search for a new New York City establishment begins. To the next one, Raihofer.

While this book was not researched nor written in chronological order, the first chapter was indeed the first chapter I wrote. Jean-Pierre Wolf and his wonderful colleague Luigi Bonacina emboldened my belief that climate modification was and is a great, great story to be told. Jean-Pierre's interview cranked the engine for all the adventures that followed.

I also want to thank everyone who took time out of their busy schedules to speak with me, and in many cases serve as a tour guide. Especially,

Jonathan Parfrey, Klaus Lackner, Jorgen Olesen, Ron Stotish, Kristian Koreman, Christopher Wan, Adam Bosch, and Jim Herberg. My most sincere gratitude.

Public relations executives, assistants, and associates, there were so many of you who paved the way for my interviews, please forgive me for not being able to name you all. But know that without your help, there would be no stories to tell.

I have been fortunate to have reported from around the world over the years, and some of those environs framed chapters in this book. However, there are several particularly memorable moments from new grounds under my feet: sitting on the terrace of a high-rise restaurant in the Dubai Marina as the sun set over the city. The most harrowing car ride of my life over the Atlas Mountains in Morocco. The near-death experience of scaling the Svartisen glacier in Norway. And the breathtaking walks through the fields in northern Denmark. Wow, just wow.

To be sure, I am leaving out so many people who assisted in the field research process, including drivers, fixers, and translators. But they are important and not forgotten.

At home, there is another army for which to be grateful. Tiffany Snow and Bre Cunningham held down the fort and took amazing care of home front logistics. My eight brothers and sisters never could remember to take me off group texts while I was overseas, so they will be receiving bills. Still, they kept "home" in mind, which is always comforting when you are holed up in a far off, strange land. For more than a decade, Jeannie Lee has answered my self-doubt with the response, "Of course you can do it." Thank you for the support and strength you give me.

Once all the research and travel and writing are done—the real work begins. Laura Mahler, my trustee researcher, you kept me honest. Sara Carder, your belief in this project exceeded any and all expectations. The same goes for Megan Newman, who "got it" the second we met. Rachel Ayotte, many thanks for keeping the book's flow going. Also, at TarcherPerigee and Penguin Random House, I'd like to give shout outs to Anne Chan, Andrea St. Aubin, Farin Schlussel, Anne Kosmoski, and Lindsay Gordon.

Finally, I would like to express my appreciation the scientific research community. These unsung heroes make the world a better place and fight the good fight sometimes in the face of opposition and ignorance. They look past polemics to solutions. I wish more of us could.

Notes

Introduction: Why Can't We?

1 **had doubled since 1980:** European Academies' Science Advisory Council, Leopoldina, "New Data Confirm Increased Frequency of Extreme Weather Events," ScienceDaily, https://www.sciencedaily.com/releases/2018/03/180321130859.htm.

1 **had jumped 50 percent in just two decades:** Lucy Cormack, "Rate of Global Sea Level Rise Jumps 50 Per Cent in Two Decades," *Sunday Morning Herald*, https://www.smh.com.au/environment/rate-of-global-sea-level-rise-jumps-50-per-cent-in-two-decades-20170626-gwyu52.html.

1 **had quadrupled from the average number forty years ago:** European Academies' Science Advisory Council, Leopoldina, "New Data Confirm Increased Frequency of Extreme Weather Events."

1 **its most destructive wildfire in modern history:** Thomas Sumner, "California Drought Worst in at Least 1,200 Years," Science News, https://www.sciencenews.org/article/california-drought-worst-least-1200-years; Jack Nicas and Thomas Fuller, "Wildfire Becomes Deadliest in California History," *New York Times*, https://www.nytimes.com/2018/11/12/us/california-fires-camp-fire.html.

1 **climate refugees by 2050:** Baher Kamal, "Climate Migrants Might Reach One Billion by 2050," Relief Web International, https://reliefweb.int/report/world/climate-migrants-might-reach-one-billion-2050.

2 **should be circulating in the air at its lowest levels:** Brian Kahn, "The World Passes 400ppm Carbon Dioxide Threshold. Permanently," *Guardian*, https://www.theguardian.com/environment/2016/sep/28/the-world-passes-400ppm-carbon-dioxide-threshold-permanently.

2 **and we broke through it:** Nicola Jones, "How the World Passed a Carbon Threshold and Why It Matters," Yale Environment 360, https://e360.yale.edu/features/how-the-world-passed-a-carbon-threshold-400ppm-and-why-it-matters.

2 **temperature rise of 5.4°F by the middle of the century:** Intergovernmental Panel on Climate Change, *Special Report: Global Warming of 1.5°C*, IPCC, https://www.ipcc.ch/sr15/.

2 **"counteract the effects of global warming":** *Oxford English Dictionary*, s.v. "Geoengineering," https://en.oxforddictionaries.com/definition/geoengineering.

2 **research for geoengineering possibilities are necessary:** National Research Council, *Climate Intervention: Carbon Dioxide Removal and Reliable Sequestration* (Washington, DC: The National Academies Press, 2015).

2 **act before unprecedented changes in all aspects of society will occur:** Intergovernmental Panel on Climate Change, "Summary for Policymakers of IPCC Special Report on Global Warming of 1.5°C Approved by Governments," IPCC, https://www.ipcc.ch/2018/10/08/summary-for-policymakers-of-ipcc-special-report-on-global-warming-of-1-5c-approved-by-governments/.

Chapter 1: Thor's Hammer

9 **from the Atlantic, Gulf Coast, and Pacific Northwest:** Mount Washington Observatory, https://www.mountwashington.org/.

11 **topping the most destructive storm season ever recorded:** "Extremely Active 2017 Atlantic Hurricane Season Finally Ends," NOAA, http://www.noaa.gov/media-release/extremely-active-2017-atlantic-hurricane-season-finally-ends; "A Look at Hurricane Irma by the Numbers," *Los Angeles Times*, September 12, 2017, http://www.latimes.com/nation/nationnow/la-na-hurricane-irma-numbers-20170912-htmlstory.html.

11 **more snow than it ever has:** Patrick Reevell, "Moscow Buried Under Record Snowfall," ABC News video, https://abcnews.go.com/International/moscow-buried-record-snowfall/story?id=52845934.

11 **The curious list of monikers goes on:** Dennis Mersereau, "The Weather Channel's Winter Storm Names Are a Cheap Advertising Ploy," The Vane, http://thevane.gawker.com/the-weather-channels-winter-storm-names-are-a-cheap-adv-1641241166.

11 **Epic, five-hundred-year floods are occurring annually:** "'500-year' Rain Events Are Happening More Often Than You Think," CBS News, https://www.cbsnews.com/news/what-does-500-year-flood-really-mean/.

11 **during just one week in July 2018:** Eleanor Rose, "All-Time Hottest Temperature Records Set All Over the World This Week," *Evening Standard*, https://www.standard.co.uk/news/world/alltime-hottest-temperature-records-set-all-over-the-world-this-week-a3881061.html.

11 **Miami and Key West, Florida, put out freeze warnings:** Diana Pearl, "The Craziest Weather Records This Seemingly Endless Cold Spell Is Breaking," People.com, https://people.com/human-interest/2018-winter-weather-records-low-temperatures/; "South Florida Rises to Record-Low Temperatures," Local10.com, https://www.local10.com/news/south-florida-rises-to-record-low-temperatures.

12 **wind speeds and precipitation have increased 5 percent:** Mason Inman, "Earth Getting Mysteriously Windier," *National Geographic*, https://news.nationalgeographic.com/news/2011/03/110328-earth-storms-winds-global-warming-science-environment/; US Global Change Research Program, chap. 7 in *Climate Science Special Report: Fourth National Climate Assessment, 2017* (Washington, DC: US Global Change Research Program, 2017), https://science2017.globalchange.gov/chapter/7/.

13 **causing flash floods that killed thirty-five people:** "Rain Making Link to Killer Floods," BBC.com, http://news.bbc.co.uk/1/hi/uk/1516880.stm.

13 **Popeye's success was uneven and eventually abandoned:** "Weather Modification in North Vietnam and Laos (Project Popeye)," Office of the Historian, https://history .state.gov/historicaldocuments/frus1964-68v28/d274.

13 **weather fix was never scientifically proven accurate:** Clifford Coonan, "How Beijing Used Rockets to Keep Opening Ceremony Dry," *Independent*, https://www .independent.co.uk/sport/olympics/how-beijing-used-rockets-to-keep-opening -ceremony-dry-890294.html.

16 **"open a number of new applications":** "Artificial Lightning: Laser Triggers Electrical Activity in Thunderstorm for the First Time," ScienceDaily, https://www.sci encedaily.com/releases/2008/04/080414082517.htm.

18 **refugee centers for food, clean water, shelter, and medical aid:** Jeffrey Gettleman, "More than 1,000 Died in South Asia Floods This Summer," *New York Times*, https:// www.nytimes.com/2017/08/29/world/asia/floods-south-asia-india-bangladesh -nepal-houston.html.

18 **are shifts in prevailing wind directions:** Rachel Ross, "What Is a Monsoon?," Live Science, https://www.livescience.com/56906-monsoon.html.

18 **or dry spells during winter:** Chloe Farand, "At Least 41 Million People Affected in Floods in India, Bangladesh and Nepal, UN Says," *Independent*, https://www .independent.co.uk/news/world/asia/india-floods-bangladesh-nepal-millions -affected-says-un-a7920721.html.

18 **an example of what could go terribly wrong:** James Fleming, *Fixing the Sky: The Checkered History of Weather and Climate Control* (New York: Columbia University Press, 2010), 160–62.

18 **HAARP was shuttered in 2014:** Stephanie Pappas, "Secret Weapon? Conspiracy Theories Abound as US Military Closes HAARP," Live Science, www.livescience .com/45829-haarp-shutdown.html.

19 **haunt weather change-makers well into the future:** Geoengineering Watch, https:// www.geoengineeringwatch.org/.

21 **glacial melt and fewer inches of snowfall each year:** Jane Qui, "Droughts Threaten High-Altitude Himalayan Forests," *Nature*, https://www.nature.com/news/droughts -threaten-high-altitude-himalayan-forests-1.16806.

22 **three times the size of Spain:** Peter Dockrill, "China's 'Sky River' Will Be the Biggest Artificial Rain Experiment Ever," Science Alert, https://www.sciencealert.com /how-china-s-sky-river-will-be-the-biggest-artificial-rain-experiment-ever-cloud -seeding.

22 **maybe even a new sky above:** Stephen Chen, "China Needs More Water. So It's Building a Rain-Making Network Three Times the Size of Spain," *South China Morning Post* (Hong Kong), https://www.scmp.com/news/china/society/article/2138866/china -needs-more-water-so-its-building-rain-making-network-three; Andrew Moseman, "Does Cloud Seeding Work?," *Scientific American*, https://www.scientificamerican .com/article/cloud-seeding-china-snow/.

22 **the cold water siphon:** Myra Per-Lee, "Far from the Computer World, Bill Gates Is Killing Hurricanes," Inventor Spot, http://inventorspot.com/articles/bill_gates _30410.

23 **educate the public and educators on marine science:** "Where Do Hurricanes Get Their Strength?" Smithsonian Ocean, https://ocean.si.edu/planet-ocean/waves -storms-tsunamis/where-do-hurricanes-get-their-strength.

23 **the approach is theoretically possible:** National Center for Atmospheric Research, "Hurricanes, Typhoons, Cyclones: Background on the Science, People, and Issues Involved in Hurricane Research," NCAR UCAR, https://news.ucar.edu/1438/hurricanes -typhoons-cyclones.

23 **or 200 miles per hour, or more:** David Fleshler, "The World Has Never Seen a Category 6 Hurricane, but the Day May Be Coming," *Ledger* (Lakeland, FL), https:// www.theledger.com/news/20180710/world-has-never-seen-category-6-hurricane -but-day-may-be-coming.
   * Jean-Pierre Wolf's most recent academic paper describes his progress in weather control, testing short lasers in the atmosphere: Jean-Pierre Wolf, "Short-Pulse Lasers for Weather Control," *Reports on Progress in Physics* 81, no.2, (2018), https://archive -ouverte.unige.ch/unige:101051.

Chapter 2: A Parasol for the Planet

24 **the weather station in Furnace Creek, California:** World Meteorological Organization's World Weather and Climate Extremes Archive, "World: Highest temperature," Arizona State University, https://wmo.asu.edu/content/world-highest -temperature.

24 **the lowest point in North America, Badwater Basin:** "Badwater Basin," National Park Service, https://www.nps.gov/places/badwater-basin.htm.

26 **and Ahvaz, Iran, in 2017:** Christopher Burt, "Hottest Reliably Measured Air Temperatures on Earth," Weather Underground, https://www.wunderground.com/blog /weatherhistorian/hottest-reliably-measured-air-temperatures-on-earth.htm.

26 **Queensland, Australia in 2003 (156.7°F):** Michael Carlowicz, "Where Is the Hottest Place on Earth? It Lies Somewhere Between Folklore and Science, the Desert and the City," NASA Earth Observatory, https://earthobservatory.nasa.gov/features /HottestSpot/page2.php.

26 **Furnace Creek's top, hot "honor":** "WMO Examines Reported Record Temperature of 54°C in Kuwait," WMO.int, https://public.wmo.int/en/media/news/wmo-examines -reported-record-temperature-of-54°C-kuwait.

26 **warmest years on record have occurred since 2001:** John Schwartz and Nadja Popovich, "It's Official: 2018 Was the Fourth-Warmest Year on Record," *New York Times*, https://www.nytimes.com/interactive/2019/02/06/climate/fourth-hottest -year.html.

26 **killed more than two thousand people:** Tom Di Liberto, "India Heat Wave Kills Thousands," Climate.gov, https://www.climate.gov/news-features/event-tracker/india -heat-wave-kills-thousands.

26 **in July 2018 due to heat illness. Six died:** "Japan Nears Record Number of Hospitalizations for Heat-Related Issues," *Japan Times* (Tokyo), https://www.japantimes .co.jp/news/2018/07/31/national/japan-nears-record-number-hospitalizations-heat -related-issues/#.XNW4PS-ZOu4.

26 **issued heat advisories and warnings:** Brandon Miller and Madison Park, "Heat Wave Turns Deadly and Is Expected to Last Through the Fourth of July," CNN.com, https://www.cnn.com/2018/07/03/us/heat-wave-wxc/index.html.

26 **several in the northeast United States:** American Meteorological Society, "Human Influence on Climate Led to Several Major Weather Extremes in 2016," Climate.gov,

https://www.climate.gov/news-features/understanding-climate/human-influence
-climate-led-several-major-weather-extremes-2016.

27 **brought about by these meteorological phenomenons:** "What Is a Heat Wave?,"
SciJinks.gov, https://scijinks.gov/heat/.

27 **"more than 4°C (7.2 °F) by the end of the 21st century":** "What Will the Weather Be
Like in 2050?," UN Sustainable Development Goals video series, https://www.un
.org/sustainabledevelopment/blog/2015/03/what-will-the-weather-be-like-in
-2050/.

28 **around the warming conundrum:** Lee Billings, "Catching the Stars," Aeon, https://
aeon.co/essays/when-this-man-talks-about-energy-the-world-needs-to-listen.

29 **in the summer of 2006:** Roger Angel and Pete Worden, "Global Warming and NA-
SA's New Vision Could Share a Common Solution Far Above the Earth," *Ad Astra*
18, no. 1 (2006), https://space.nss.org/making-sun-shades-from-moondust/.

31 **power all the world's energy needs for a full year:** Stephen Cass, "Solar Power Will
Make a Difference—Eventually," *MIT Technology Review*, https://www.technology
review.com/s/414792/solar-power-will-make-a-difference-eventually/.

31 **attributed to human-caused global warming:** Findings were later published in
D. Lunt et al., "'Sunshade World': A Fully Coupled GCM Evaluation of the Climatic
Impacts of Geoengineering," *Geophysical Research Letters* 35, no. 12 (June 2008).

31 **between 180 parts per million and 290 parts per million:** Nicola Jones, "How the
World Passed a Carbon Threshold and Why it Matters," Yale Environment 360,
https://e360.yale.edu/features/how-the-world-passed-a-carbon-threshold-400ppm
-and-why-it-matters.

32 **the convection cycle continues:** Jürgen Schieber, "Global Energy Transfer, Atmosphere, Climate," chap. 4 in *Earth: Our Habitable Planet* (Indiana University, 2019),
5 http://www.indiana.edu/~geol105/1425chap4.htm.

33 **uses 3-D printing methods to make equipment:** "International Space Station's
3-D printer," NASA, https://www.nasa.gov/content/international-space-station-s-3
-d-printer.

33 **to create entirely new structures:** Made In Space, https://madeinspace.us/.

34 **at a rate not seen since the time of dinosaurs:** Matt McGrath, "Nature Crisis: Humans 'Threaten 1m Species with Extinction,'" BBC.com, https://www.bbc.co.uk
/news/science-environment-48169783.

34 **nine million different types of species on Earth:** Census of Marine Life, "How
Many Species on Earth? About 8.7 Million, New Estimate Says," ScienceDaily,
https://www.sciencedaily.com/releases/2011/08/110823180459.htm.

34 **half of all species on the planet may be gone by mid-century:** "The Extinction
Crisis," Center for Biological Diversity, https://www.biologicaldiversity.org/pro
grams/biodiversity/elements_of_biodiversity/extinction_crisis/.

35 **away from the sun to cool the planet:** Zaria Gorvett, "How a Giant Space Umbrella
Could Stop Global Warming," BBC Future, http://www.bbc.com/future/story
/20160425-how-a-giant-space-umbrella-could-stop-global-warming.

35 **to study Jupiter and Saturn:** "Voyager 2," NASA Science, https://solarsystem.nasa
.gov/missions/voyager-2/in-depth/.

36 **landed a drone on a comet, proving there is a way:** Tom Risen, "Rosetta Comet
Landing Is Space Game-Changer," U.S. News & World Report, https://www.usnews
.com/news/articles/2014/11/12/rosetta-comet-landing-is-space-game-changer.

36 **the only thing left of our biosphere would likely be bacteria:** Robin McKie, "NASA's Answer to Global Warming: Move the Earth," Rense.com, https://rense.com/general11/move.htm.

37 **But for the time being, we'd stay cool:** Neil Comins, "What if the Moon Didn't Exist?," *Astronomical Society of the Pacific* 33 (Winter 1996), https://astrosociety.org/edu/publications/tnl/33/33.html.

37 **"doesn't seem immediately feasible":** Mark Williams Pontin, "Cooling the Planet," *MIT Technology Review*, https://www.technologyreview.com/s/407306/cooling-the-planet/.

38 **by 2022 for just ten billion dollars:** Fiona MacDonald, "NASA Scientists Say We Could Colonise the Moon by 2022 . . . for Just $10 Billion," Science Alert, https://www.sciencealert.com/nasa-scientists-say-we-could-colonise-the-moon-by-2022-for-just-10-billion.

38 **more than seventy times that amount:** Jeff Stein and Aaron Gregg, "US Military Spending Set to Increase for Fifth Consecutive Year, Nearing Levels During Height of Iraq War," *Washington Post*, https://www.washingtonpost.com/us-policy/2019/04/18/us-military-spending-set-increase-fifth-consecutive-year-nearing-levels-during-height-iraq-war/?noredirect=on&utm_term=.0909813536b9.

Chapter 3: Carbon Vampires

40 **using the world's most advanced seismic sensors:** Bec Crew, "Massive Amounts of Melting Carbon Have Been Found Under the Western US," Science Alert, https://www.sciencealert.com/massive-amounts-of-melting-carbon-have-been-found-under-the-western-us.

40 **emit in a year:** "Green Vehicle Guide," US Environmental Protection Agency, https://www.epa.gov/greenvehicles/greenhouse-gas-emissions-typical-passenger-vehicle.

41 **will be done on the area, he assures:** Sean Madden, "Yellowstone Supervolcano Due for an Eruption," The Weather Network, https://www.theweathernetwork.com/news/articles/yellowstone-volcano-overdue-for-an-eruption-super-eruption/99739; Mike McRae, "Cracks Are Appearing in the Rocks Near Yellowstone. Here's Why You Don't Need to Panic," Science Alert, https://www.sciencealert.com/grand-teton-national-park-yellowstone-rock-fissures-close-hidden-falls.

41 **amounts to around forty billion tons:** Scott Michon and Rebecca Lindsey, "Which Emits More Carbon Dioxide: Volcanoes or Human Activities?," Climate.gov, https://www.climate.gov/news-features/climate-qa/which-emits-more-carbon-dioxide-volcanoes-or-human-activities.

41 **cut by about 25 percent—right now:** "Report: World Must Cut Further 25% from Predicted 2030 Emissions," UN Sustainable Development Goals, https://www.un.org/sustainabledevelopment/blog/2016/11/report-world-must-cut-further-25-from-predicted-2030-emissions/.

41 **50 percent by 2030:** Brad Plumer and Nadja Popovich, "The World Still Isn't Meeting Its Climate Goals," *New York Times*, https://www.nytimes.com/interactive/2018/12/07/climate/world-emissions-paris-goals-not-on-track.html.

42 **have risen nearly 3.6°F (2°C) over the same period:** University of California, San Diego, "Climate Change: Climate in the Spotlight," Earth Guide UCSD, 2002, http://earthguide.ucsd.edu/virtualmuseum/climatechange1/01_1.shtml.

42 **carbon dioxide produced by humans is absorbed:** "Oceans and the Carbon Cycle," GDRC.org, http://www.gdrc.org/oceans/fsheet-02.html.

42 **25 percent of human-produced carbon dioxide:** "Ocean-Atmosphere CO_2 Exchange," NOAA Science on a Sphere, https://sos.noaa.gov/datasets/ocean-atmo sphere-co2-exchange/.

42 **one ton of carbon from the atmosphere:** Erv Evans, "Tree Facts," North Carolina State University College of Agriculture and Life Sciences, https://projects.ncsu.edu /project/treesofstrength/treefact.htm.

42 **three Indias, according to some estimates:** Renee Cho, "Can Removing Carbon from the Atmosphere Save Us from Climate Catastrophe?," Phys.org, https://phys .org/news/2018-11-carbon-atmosphere-climate-catastrophe.html.

45 **roots rather than a metallic tank:** University of Helsinki, "Carbon Tree," Carbon Cycle, http://hiilipuu.fi/articles/carbon-cycle.

45–46 **the possibility isn't all that far-fetched:** Rachel Ehrenberg, "Global Count Reaches 3 Trillion Trees," *Nature,* https://www.nature.com/news/global-count-reaches-3 -trillion-trees-1.18287.

* This map is an illustration of recent global afforestation patterns: "Global Planted Forests 1990–2015," Carbon Brief, https://www.carbonbrief.org/mapped-where -afforestation-is-taking-place-around-the-world.

46 **more than a billion automobiles in the world:** John Voelcker, "1.2 Billion Vehicles on World's Roads Now, 2 Billion by 2035: Report," Green Car Reports, https:// www2.greencarreports.com/news/1093560_1-2-billion-vehicles-on-worlds-roads -now-2-billion-by-2035-report.

46 **forty million shipping containers filled with stuff per year:** "Shanghai Port Sets World Box Volume Record," World Maritime News, https://worldmaritimenews .com/archives/239541/shanghai-port-sets-new-world-box-volume-record.

48 **"make climate change worse and threaten people and animals":** Greenpeace USA, *Carbon Capture SCAM: How a False Climate Solution Bolsters Big Oil* (Washington, DC: Greenpeace, April 15, 2015), https://senate.ucsd.edu/media/206094/carbon -capture.pdf.

49 **gas at the pump, electric charging, or hydrogen filling:** James Temple, "Maybe We Can Afford to Suck CO_2 Out of the Sky After All," *MIT Technology Review,* https:// www.technologyreview.com/s/611369/maybe-we-can-afford-to-suck-cosub2sub -out-of-the-sky-after-all/.

49 **"to supply to customers," Climeworks explains:** Climeworks, http://www .climeworks.com.

49 **forming solid minerals:** Akshat Rathi, "The World's First 'Negative Emissions' Plant Has Begun Operation—Turning Carbon Dioxide into Stone," Quartz, https:// qz.com/1100221/the-worlds-first-negative-emissions-plant-has-opened-in-iceland -turning-carbon-dioxide-into-stone/.

50 **which is in line with gas prices at the pump:** Robert Service, "Cost Plunges for Capturing Carbon Dioxide from the Air," *Science,* http://www.sciencemag.org/news /2018/06/cost-plunges-capturing-carbon-dioxide-air.

51 **various designs for using their carbon-storage potential:** New York Times News Service, "How Oman's Rocks Could Help Save the Planet," Gulf News, https:// gulfnews.com/world/gulf/oman/how-omans-rocks-could-help-save-the -planet-1.2213007.

51 **"per cubic kilometer of rock per year," the geologists estimate:** Lamont-Doherty Earth Observatory, "Carbon Sequestration," Columbia University Earth Institute, https://www.ldeo.columbia.edu/gpg/projects/carbon-sequestration.

51 **capture very little of that back through human ingenuity:** Caleb Scharf, "The Crazy Scale of Human Carbon Emission," *Scientific American*, https://blogs.scientific american.com/life-unbounded/the-crazy-scale-of-human-carbon-emission/.

51 **that statistic is on the very high side:** Ross Gorte, *CRS Report R40562: US Tree Planting for Carbon Sequestration* (Washington, DC: Congressional Research Service, May 4, 2009), https://fas.org/sgp/crs/misc/R40562.pdf.
* Shortly after the writing of this chapter, Klaus Lackner and his team at Arizona State University partnered with Silicon Kingdom Holdings, a Dublin-based tech start-up, to commercialize and deploy his trees: Skip Derra, "Lackner's Carbon -Capture Technology Moves to Commercialization," ASU Now, https://asunow.asu .edu/20190429-solutions-lackner-carbon-capture-technology-moves-commercial ization.

Chapter 4: Desert Reflectors

52 **the most in one concentrated swath:** "Sustainable Development Goal 7," International Energy Agency, https://www.iea.org/sdg/electricity/.

53 **world's needs more than eighty times over:** Mehran Moalem, "We Could Power the Entire World by Harnessing Solar Energy from 1.2% of the Sahara," *Forbes*, https:// www.forbes.com/sites/quora/2016/09/22/we-could-power-the-entire-world -by-harnessing-solar-energy-from-1-of-the-sahara/#2d7d7fc3d440.

53 **from Mauritania to South Africa:** "Sub-Saharan Africa—World Bank Open Data," World Bank Group, https://data.worldbank.org/region/sub-saharan-africa.

54 **do not have connections to electrical power:** "Sustainable Development Goal 7," International Energy Agency.

54 **can't plug in and power up:** Ibid.

54 **widely use fossil fuels as the source:** Todd Lindeman, "1.3 Billion People Are Living in the Dark," *Washington Post*, https://www.washingtonpost.com/graphics/world /world-without-power/??noredirect=on.

54 **from the burning of coal and other fossil-related fuels:** "Electricity Explained: Electricity and the Environment," US Energy Information Administration, https:// www.eia.gov/energyexplained/index.php?page=electricity_environment.

54 **Globally, it's about the same:** "World Energy Council Report Confirms Global Abundance of Energy Resources and Exposes Myth of Peak Oil," World Energy Council, https://www.worldenergy.org/news-and-media/press-releases/world-energy -council-report-confirms-global-abundance-of-energy-resources-and-exposes -myth-of-peak-oil/.

54 **adding further to energy demands there:** "Future World Energy Demand Driven by Trends in Developing Countries," US Energy Information Administration, https://www.eia.gov/todayinenergy/detail.php?id=14011.

55 **ten billion lightbulbs of the one-hundred-watt variety:** "Current World Energy Consumption," The World Counts, http://www.theworldcounts.com/stories/current _world_energy_consumption.

55 **will increase by nearly 30 percent:** "EIA Projects 28% Increase in World Energy Use by 2040," US Energy Information Administration, https://www.eia.gov/today inenergy/detail.php?id=32912.

55 **four times that amount—up 124 percent:** Jason Deign, "Global Energy Demand Could Grow 124% by 2100: Even Fossil Fuels Won't Cut It," Green Tech Media, https://www.greentechmedia.com/articles/read/energy-needs-in-2100-even-fossil -fuels-will-not-cut-it#gs.7in8fo.

55 **1.5 percent of the world's energy supply:** Hannah Ritchie and Max Roser, "Renewable Energy," Our World in Data, https://ourworldindata.org/renewable-energy.

56 **40 percent of the land area on Earth:** Christina Nunez, "Deserts, Explained," *National Geographic*, https://www.nationalgeographic.com/environment/habitats /deserts/.

56 **two thousand miles away from a major desert area:** "The Vision," Desertec, https:// www.desertec.org/.

56 **pressure that forces air into motion:** Chris Weiss, "Where Does Wind Come From?" *Scientific American*, https://www.scientificamerican.com/article/where-does-wind -come-from/.

56 **studies show that mood shifts do occur:** E. H. Bos, R. Hoenders, and P. de Jonge, "Wind Direction and Mental Health: A Time-Series Analysis of Weather Influences in a Patient with Anxiety Disorder," *BMJ Case Reports 2012*, June 2012; Alfredo Llorente, "How Does Weather Affect Crime Rates?," City Data, http://www.city-data .com/blog/28-weather-affect-crime-rates/.

57 **is less than one-third the size:** "The World's Largest Deserts," Geology.com, https:// geology.com/records/largest-desert.shtml.

58 **Middle East/North Africa (MENA) region and Europe remains:** Sonja van Renssen, "EU's Mediterranean Neighbours Struggle with Energy Transition," Energy Central, https://www.energycentral.com/c/ec/eus-mediterranean-neighbours -struggle-energy-transition.

58 **what people pay on average in Europe:** "Electricity Prices in Europe—Who Pays the Most?," *Strom-Report Blog*, Stromvergleich, https://1-stromvergleich.com /electricity-prices-europe/. Based on Euro cents.

59 **expand Europe's colonial reach:** Susan Fourtané, "Atlantropa: Herman Sörgel's Vision of a New Continent," Intereresting Engineering, https://interestingengineering .com/atlantropa-herman-sorgels-vision-of-a-new-continent.

59 **by deep-sea power cables:** "Tunisian Solar Power to Reach Europe via Three Subsea Cables," Subsea World News, https://subseaworldnews.com/2017/08/11/tunisian -solar-power-to-reach-europe-via-three-subsea-cables/.

62 **article for the scientific journal *Nature Climate Change*:** Aixue Hu et al., "Impact of Solar Panels on Global Climate," *Nature Climate Change* 6 (2016): 290–94.

64 **generates electricity via an alternator:** Liz Stinson, "The Future of Wind Turbines? No Blades," *Wired*, https://www.wired.com/2015/05/future-wind-turbines -no-blades/.

64 **to spin at a particular point:** "Vorticity," University of Waterloo, https://uwaterloo .ca/applied-mathematics/current-undergraduates/continuum-and-fluid-mechanics -students/amath-463-students/vorticity.

64 **the company explains:** Vortex Bladeless, https://vortexbladeless.com/technology -design/.

65 **approximately three hundred feet high:** "FAQ—Size," National Wind Watch, https://www.wind-watch.org/faq-size.php.

66 **could power the needs for all of humankind:** A. Possner and K. Caldeira, "Geophysical Potential for Wind Energy over the Open Oceans," *Proceedings of the National Academy of Sciences of the United States of America* 114, no. 43 (October 2017): 11338–43.

66 **It just acts as a conduit:** Chris Woodford, "OTEC—Ocean Thermal Energy Conversion," Explain That Stuff!, https://www.explainthatstuff.com/how-otec-works.html.

67 **began operating in Hawaii in 2015:** Makai Ocean Engineering, http://www.makai.com/makai-news/2015_08_29_makai_connects_otec/.

67 **the future of what ocean energy might look like:** Okinawa OTEC, http://otecokinawa.com/en/index.html.

67 **from Asia to the US East Coast:** Ocean Thermal Energy Corporation, http://otecorporation.com/.

67 **several different types of wave-energy devices:** "Wave Energy," Alternative Energy Tutorials, http://www.alternative-energy-tutorials.com/wave-energy/wave-energy.html.

67 **wave power is on the rise:** Borja Reguero et al., "A Recent Increase in Global Wave Power as a Consequence of Oceanic Warming," *Nature Communications* 10 (2019), https://doi.org/10.1038/s41467-018-08066-0.

Chapter 5: Cool Roofs and Roads

68 **dark surrounding desert terrain at night:** NASA, "Las Vegas at Night," NASA Visible Earth, https://visibleearth.nasa.gov/view.php?id=47687.

68 **hotels and casinos illuminate so magnificently:** "How Much Electricity Does Las Vegas Use per Day?," Reference, https://www.reference.com/geography/much-electricity-las-vegas-use-per-day-b455912581f24e4b.

68 **thirty-nine xenon lamps:** Hugh Morris, "The Casino Light Beam That's So Bright It Has Its Own Ecosystem (and Pilots Use It to Navigate)," *Telegraph* "Travel," https://www.telegraph.co.uk/travel/travel-truths/famous-waypoints-aviation-luxor-sky-beam/.

69 **242 days of clear skies per year:** Liz Osborn, "Sunniest Places in the United States," Current Results, https://www.currentresults.com/Weather-Extremes/US/sunniest.php.

69 **appear most to the naked eye:** Yuma, Arizona, World Population Review, http://worldpopulationreview.com/us-cities/yuma-population/.

69 **dark colors creating artificial hot zones:** "Cooler Cities," Global Cool Cities Alliance, https://globalcoolcities.org/discover/cool-science/cooler-cities/.

69 **expected to triple in size by 2030:** David Biello, "Gigalopolises: Urban Land Area May Triple by 2030," *Scientific American*, https://www.scientificamerican.com/article/cities-may-triple-in-size-by-2030/.

70 **asthma, heart attacks, strokes, lung cancer, and pre-term births:** Tony Barboza and Jon Schleuss, "LA Keeps Building Near Freeways, Even Though Living There Makes People Sick," *Los Angeles Times*, https://www.latimes.com/projects/la-me-freeway-pollution/.

70 **twenty degrees hotter than its surrounding suburbs:** Linda Poon, "Street Grids Make Cities Hotter," City Lab, https://www.citylab.com/environment/2018/04 /street-grids-may-make-cities-hotter/558845.

71 **at residents of New York:** David Wallace-Wells, "The Uninhabitable Earth, Annotated Edition," *New York*, http://nymag.com/intelligencer/2017/07/climate -change-earth-too-hot-for-humans.html.

71 **all new homes use roof materials that reflect sunlight:** "Cool Roofs and Surfaces," Climate Resolve, https://climateresolve.org/our-work/keeping-la-cool/cool-surfaces/. * As of this writing, Climate Resolve has helped to cool 27,000 roofs across Los Angeles: "Our Impact," Climate Resolve, https://www.climateresolve.org/.

71 **reduce the heat temperature of the city by 3°F by 2035:** Nate Berg, "The Radical Plan to Cool Down LA as the World Heats Up," Gizmodo, https://gizmodo.com/the -radical-plan-to-cool-down-la-as-the-world-heats-up-1797711611.

71 **expected to jump 3.6°F by that time:** Laurie L. Dove, "Study Says 2035 Is Climate Change Point of No Return," HowStuffWorks.com, https://science.howstuffworks .com/environmental/conservation/issues/point-no-return-for-climate-action -is-2035.htm.

72 **And that's just water:** P. Vahmani and A. Jones, "Water Conservation Benefits of Urban Heat Migration," *Nature Communications* 8 (2017): 1072.

72 **the size of Los Angeles or greater:** Daniel Maina Wambugu, "Largest Cities in the World By Land Area," World Atlas, https://www.worldatlas.com/articles/largest -cities-in-the-world-by-land-area.html.

72 **expected to triple by mid-century:** A. Peters, "This Is How Hot Your City Could Be by the End of the Century," *Fast Company*, https://www.fastcompany.com/40439265 /this-is-how-hot-your-city-could-be-by-the-end-of-the-century.

72 **due to medical care, labor, and other losses:** Alister Doyle, "Urban 'Heat Islands' Seen Doubling City Costs for Climate Change," Reuters, https://www.reuters.com /article/us-climatechange-cities-idUSKBN18P1KS.

73 **cooler than, say, a darker one:** Arthur Rosenfeld, "White Roofs Cool the World Efficiently," International Energy Agency, https://www.iea.org/newsroom/news/2013 /april/white-roofs-cool-the-world-efficiently.html.

74 **if all US cities lightened their loads:** "Cool Roofs," Green Building Alliance, https:// www.go-gba.org/resources/green-building-methods/cool-roofs/.

75 **part of the Earth's own thermal emission:** Columbia University, "The Climate System: Solar Radiation, Earth's Atmosphere, and the Greenhouse Effect," Department of Earth and Environmental Sciences, https://eesc.columbia.edu/courses/ees /climate/lectures/radiation_hays; "Earth's Energy Budget," NASA Earth Observatory, January 14, 2009, https://earthobservatory.nasa.gov/features/EnergyBalance /page4.php.

76 **rests at about 59°F:** Tim Sharp, "What Is Earth's Average Temperature?," Space .com, https://www.space.com/17816-earth-temperature.html.

77 **the Earth overall is warmed:** Mark Jacobson, "Effects of Urban Surfaces and White Roofs on Global and Regional Climate," Stanford University Department of Civil and Environmental Engineering, https://web.stanford.edu/group/efmh/jacobson /Articles/Others/HeatIsland+WhiteRfs0911.pdf.

78 **smog has been around for centuries:** Online Etymology Dictionary, s.v. "smog," https://www.etymonline.com/word/smog.

79 **and lung cancer can also be caused:** "Nitrogen Dioxide (NO2) Pollution," US Environmental Protection Agency, https://www.epa.gov/no2-pollution/basic-information -about-no2#What%20is%20NO2.

79 **where air quality exceeds World Health Organization standards:** "Air Pollution," World Health Organization, https://www.who.int/airpollution/en/.

79 **"improving air quality," 3M explains:** "3M Smog-Reducing Granules Harness the Power of the Sun to Improve Air Quality," 3M, https://news.3m.com/pt/node/6313.

80 **than any other city in the world:** Liz McEnaney, "Green Alleys: Servicing the Future," Build a Better Burb, http://buildabetterburb.org/green-alleys-servicing -the-future/.

80 **making it a famous urban heat island:** Marlene Cimons, "The Shape of Your City Could Determine How Hot It Gets at Night," *Popular Science*, https://www.popsci .com/city-shape-grid-heat-island-temperature.

80 **has launched a Green Alley Program:** Richard Daley, "The Chicago Green Handbook,"Chicago.gov,https://www.cityofchicago.org/content/dam/city/depts/cdot/Green _Alley_Handbook_2010.pdf.

80 **and compromises public health:** "Heat Island Impacts," US Environmental Protection Agency, https://www.epa.gov/heat-islands/heat-island-impacts.

80 **because of public health costs and labor losses:** Francisco Estrada et al., "A Global Economic Assessment of City Policies to Reduce Climate Change Impacts," *Nature Climate Change* 7 (2017): 403–8.

81 **on 237 square miles that is a grid system:** Chicago, World Population Review, http:// worldpopulationreview.com/us-cities/chicago-population/; City of Chicago, "Facts and Statistics," Chicago.gov, https://www.chicago.gov/city/en/about/facts.html.

81 **health effects on humans (disrupting our circadian rhythms):** "Human Health," International Dark-Sky Association, http://darksky.org/light-pollution/human -health/.

Chapter 6: Smart Soils

85 **largest reserve of tropical peat carbon is stored:** X. Comas et al., "Imaging Tropical Peatlands in Indonesia Using Ground Penetrating Radar (GPR) and Electrical Resistivity Imaging (ERI): Implications for Carbon Stock Estimates and Peat Soil Characterization," *Biogeosciences* 12 (2015): 191–229, https://www.cifor.org/library/5464/.
  * Areas of primary and peat forest in Indonesia are set to be better protected by the end of 2019, under a new permanent ban: Hans Nicholas, "Indonesian Ban on Clearing New Swaths of Forest to Be Made Permanent," Mongabay, https://news.mong abay.com/2019/06/indonesian-ban-on-clearing-new-swaths-of-forest-to-be-made -permanent/.

85 **two years' worth of fossil fuels all at once:** "World Energy Resources: Peat," World Energy Council, https://www.worldenergy.org/wp-content/uploads/2013/10/WER _2013_6_Peat.pdf.

85 **nine thousand tons of carbon per hectare:** "Indonesia's Peat Fires Add to Global Warming," *Straits Times* (Singapore), https://www.straitstimes.com/asia/se-asia /indonesias-peat-fires-add-to-global-warming.

85 **spread out thick in its tropical rainforest:** Sustainable Management of Peatland Forests in Southeast Asia, "Indonesia," ASEAN Peat, http://www.aseanpeat.net

/index.cfm?&menuid=68. Gives 13 million hectares with total carbon storage of up to 124.2 gigatons.

85 **as well as man-made structures:** Natural Resources Conservation Service New York, "What Soil Does," US Department of Agriculture, https://www.nrcs.usda .gov/wps/portal/nrcs/detail/ny/soils/health/?cid=nrcseprd1197209.

86 **twenty-eight hundred years to sequester in the ground:** S. Kurnianto et al., "Carbon Accumulation of Tropical Peatlands over Millennia: A Modelling Approach," *Global Change Biology* 21, no. 1 (2015): 431–44.

86 **sixteen million tons of carbon dioxide every day:** World Bank Group, "Indonesia's Fire and Haze Crisis," The World Bank, http://www.worldbank.org/en/news /feature/2015/12/01/indonesias-fire-and-haze-crisis.

86 **five tons of CO_2 in an entire year:** "Green Vehicle Guide," US Environmental Protection Agency, https://www.epa.gov/greenvehicles/greenhouse-gas-emissions -typical-passenger-vehicle.

86 **"responsible for other environmental and social harm":** Greenpeace International, *How the Palm Oil Industry Is Still Cooking the Climate*, (Amsterdam: Greenpeace, November 2017), http://www.greenpeace.org/archive-international/Global /international/publications/forests/2017/Still-Cooking-the-Climate.pdf.

88 **desertification has taken its toll on the island:** "Desertification, Drought, and Land: Iceland's National Report," UN Sustainable Development Goals, https:// sustainabledevelopment.un.org/index.php?page=view&nr=439&type=504 &menu=139.

88 **What were they thinking?:** Jared Diamond, *Collapse* (New York: Viking Press, 2005).

89 **just three centimeters of healthy, new top soil:** Chris Arsenault, "Only 60 Years of Farming Left if Soil Degradation Continues," *Scientific American*, https://www .scientificamerican.com/article/only-60-years-of-farming-left-if-soil-degradation -continues/.

90 **our current food supply quickly disappears:** "Why Soil Organic Matter Is So Important," Sustainable Agriculture Research and Education, https://www.sare.org /Learning-Center/Books/Building-Soils-for-Better-Crops-3rd-Edition/Text -Version/Organic-Matter-What-It-Is-and-Why-It-s-So-Important/Why-Soil -Organic-Matter-Is-So-Important.

91 **"death of the microbial population and thus of the soil itself":** Alexandra Bot and José Benites, "Practices that Influence the Amount of Organic Matter," in *The Importance of Soil Organic Matter: Key to Drought-Resistant Soil and Sustained Food Production*, FAO Soils Bulletin 80 (2005), http://www.fao.org/3/a0100e/a0100e07 .htm#bm07.1.

91 **methods are spreading throughout the world:** "SmartSOIL Toolbox," Ecologic Institute, https://www.ecologic.eu/12747.

92 **major contributor to ozone layer depletion:** Peter Grace and Louise Barton, "Meet N20, the Greenhouse Gas 300 Times Worse than CO_2," The Conversation, http:// theconversation.com/meet-n2o-the-greenhouse-gas-300-times-worse-than-co2 -35204.

92 **"hold liquids, gases and biota and support plants":** "Glossary of Soil Science Terms: Soil," Soil Science Society of America, https://www.soils.org/publications/soils -glossary.

93 **from modern-day Turkey to Egypt:** Ann Gibbons, "The World's First Farmers Were Surprisingly Diverse," *Science*, https://www.sciencemag.org/news/2016/07/worlds -first-farmers-were-surprisingly-diverse.

93 **The agricultural revolution began:** Carl Zimmer, "How the First Farmers Changed History," *New York Times*, https://www.nytimes.com/2016/10/18/science/ancient -farmers-archaeology-dna.html.

93 **Agricultural commodities topped global trade:** xandrewwatson, "A Quick Exploration of Ten Nineteenth Century British Imports," Trading Consequences, http:// tradingconsequences.blogs.edina.ac.uk/2014/04/08/a-quick-exploration-of-ten -nineteenth-century-british-imports/.

95 **glass and steel and new parts:** "The Apple Data Center FAQ," Data Center Knowledge, https://www.datacenterknowledge.com/data-center-faqs/apple-data-center-faq.

95 **where large, industrial farming thrives, is 443 acres:** "Average Farm Size in the United States from 2000 to 2018 (in Acres)," Statista, https://www.statista.com /statistics/196106/average-size-of-farms-in-the-us-since-2000/.

95 **more than three tons per acre per season:** Greg Blonde, "Buying and Selling Corn Silage: What's a Fair Price?," UW Extension, https://fyi.extension.wisc.edu/forage /files/2014/01/BuyingSellingCS.pdf.

95 **using dams and canals:** Irrigation Association, "Irrigation Timeline," Irrigation Museum, http://www.irrigationmuseum.org/exhibit2.aspx.

96 **assessed using computer models:** BioCarbon Engineering, https://www.biocarbon engineering.com/.

96 **land on Earth is farmland:** James Owen, "Farming Claims Almost Half Earth's Land, New Maps Show," *National Geographic*, https://news.nationalgeographic .com/news/2005/12/agriculture-food-crops-land/.

96 **land was used for food production:** Max Roser and Hannah Ritchie, "Land Use," Our World in Data, https://ourworldindata.org/land-use.

96 **green forest and vegetation growing wild:** Ibid.

96 **intelligent solutions in the field:** Agrointelli, http://www.agrointelli.com/.

98 **birth defects, cancer, and deaths:** "Birth Defects Caused by Herbicides, Insecticides, and Industrial Chemicals Disrupt the Endocrine System," Alexander Law Group LLP, https://www.alexanderlaw.com/library-toxic-3.

99 **most are small and family run:** S. Lowder, J. Skoet, and T. Raney, "The Number, Size and Distribution of Farms, Smallholder Farms, and Family Farms Worldwide," *World Development* 87 (2016): 16–29, doi.org/10.1016/j.worlddev.2015.10.041.

99 **vertical farms can be found in urban environments:** "Vertical Farm," Maximum Yield, https://www.maximumyield.com/definition/2191/vertical-farm.

99 **without sunlight or soil:** AeroFarms, https://aerofarms.com/technology/.

100 **little more than thirty-six thousand pounds:** "Lettuce," Agricultural Marketing Resource Center, https://www.agmrc.org/commodities-products/vegetables/lettuce.

100 **maximum production—indoors and out of sight:** US Department of Agriculture, "Hydroponics," National Agriculture Library, https://www.nal.usda.gov/afsic/hydro ponics.

101 **as far back as the thirteenth century:** "A Brief History of Modern Farming," Bowery Farming News, https://news.boweryfarming.com/https-medium-com -boweryfarming-a-brief-history-bda861d333d3.

101 **operates underground in an old air-raid shelter:** Zlata Rodionova, "Inside London's First Underground Farm," *Independent,* https://www.independent.co.uk /Business/indyventure/growing-underground-london-farm-food-waste-first-food -miles-a7562151.html.

101 **higher-yielding environments vertical farms can offer:** Euan McKirdy, "The Only Way Is Up: Vertical Farming in Kyoto," CNN.com, https://edition.cnn.com/travel /article/kyoto-vertical-farm-spread/index.html.

101 **thirty million dollars to build out:** Malavika Vyawahare, "World's Largest Vertical Farm Grows Without Soil, Sunlight or Water in Newark," *Guardian,* https://www .theguardian.com/environment/2016/aug/14/world-largest-vertical-farm-newark -green-revolution.

102 **countries such as the United States:** Scott Elliott, "Global Scientists Meet for Integrated Pest Management Idea Sharing," US Department of Agriculture, https://nifa .usda.gov/blog/global-scientists-meet-integrated-pest-management-idea-sharing.

103 **detailed information about every square foot of cropland:** AgEagle, https://www .ageagle.com/analytics.

103 **equipped with multispectral sensors and specialized cameras:** senseFly, https:// www.sensefly.com/.

103 **and revive lost land:** Anthony King, "Technology: The Future of Agriculture," *Nature* 544 (2017): S21–S23, doi.org/10.1038/544S21a.

* A note on vertical farming: though the costs and energy needs are high, market interest and investment is going up rapidly around the world and there are a lot of schemes trying to make it more energy and resource efficient.

Chapter 7: Brightening Clouds

104 **large bodies of water heated by the sun:** "Jumeirah Beach Water Temperature and Wetsuit Guide: Persian Gulf, United Arab Emirates," Surf-Forecast.com, http:// www.surf-forecast.com/breaks/Jumeirah-Beach-Dubai/seatemp.

104 **about 30 degrees cooler:** "State of the Climate: Global Climate Report for July 2018," NOAA National Centers for Environmental Information, https://www.ncdc.noaa .gov/sotc/global/201807.

105 **desalination process is dense with salt:** "Dispute over Cause of Persian Gulf Salinity," *Financial Tribune* (Tehran), https://financialtribune.com/articles/people -environment/43493/dispute-over-cause-of-persian-gulf-salinity.

105 **warmer global sea temperatures and higher salt levels:** University of British Columbia, "Climate Change to Cause Dramatic Drop in Persian Gulf Biodiversity and Fisheries Potential," Phys.org, https://phys.org/news/2018-05-climate-persian -gulf-biodiversity-fisheries.html.

105 **stretching far out into the Arabian Sea:** Alexandru Micu, "World's Largest Dead Zone Identified in the Gulf of Oman—It's Nearly the Size of Florida," ZME Science, https://www.zmescience.com/science/oman-gulf-dead-zone-2924534/.

105 **acclimatize to higher temperatures:** Jonah Mandel, "In the Red Sea, Coral Reefs Can Take the Heat of Climate Change," Phys.org, https://phys.org/news/2017-06-red -sea-coral-reefs-climate.html.

105–106 **continue into the foreseeable future:** "Ocean Warming," International Union for Conservation of Nature, https://www.iucn.org/resources/issues-briefs/ocean -warming.

106 **like all animals, they produce waste:** "Are Corals Animals or Plants?," NOAA National Ocean Service, https://oceanservice.noaa.gov/facts/coral.html; "Corals Get Their Food from Algae Living in Their Tissues or by Capturing and Digesting Prey," NOAA Florida Keys National Marine Sanctuary, https://floridakeys.noaa.gov /corals/coralseat.html.

106 **can be seen through the clear polyp bodies:** Coral Reef Alliance, "Coral Polyps—Tiny Builders," Coral.org, https://coral.org/coral-reefs-101/coral-reef-ecology/coral-polyps/.

106 **food into waste that algae photosynthesizes:** "Are Corals Animals or Plants?," NOAA National Ocean Service.

106 **what is known as coral bleaching:** "What Is Coral Bleaching?," NOAA National Ocean Service, https://oceanservice.noaa.gov/facts/coral_bleach.html.

107 **since record-keeping began in the 1880s:** Terry Hughes, "Global Warming and Repeated Bouts of Coral Bleaching," Nature Research Ecology and Evolution, https:// natureecoevocommunity.nature.com/users/31192-terry-hughes/posts/15551 -global-warming-and-repeated-bouts-of-coral-bleaching.

107 **threatening their existence:** "Coral Bleaching During & Since the 2014–2017 Global Coral Bleaching Event," NOAA Coral Reef Watch, https://coralreefwatch.noaa.gov /satellite/analyses_guidance/global_coral_bleaching_2014-17_status.php.

107 **reefs are important to preserve:** James Tidwell and Geoff Allan, "Fish as Food: Aquaculture's Contribution," *EMBO Reports* 2, no. 11 (2001): 958–63, doi: 10.1093 /embo-reports/kve236.

107 **"opened to the public in 2018," *National Geographic* reported:** Jenna Belhumeur and Elena Boffetta, "Why This Country Is Moving Its Coral Reefs," *National Geographic*, https://www.nationalgeographic.com/travel/destinations/asia/jordan/aqaba -coral-reef-relocation-scuba-activities/.

107 **can be easily spotted from outer space:** "The Facts," Great Barrier Reef Foundation, https://www.barrierreef.org/the-reef/the-facts.

108 **jeopardizing big sections of the reef:** Josh Gabbatiss, "Australian Government Announces £35m Plan to Save Great Barrier Reef," *Independent*, https://www .independent.co.uk/environment/great-barrier-reef-australia-plan-save-coral -bleaching-a8172986.html.

108 **gathering weather and water-quality measurements:** Melissa Matthews, "Surfing Robots Are Now Patrolling the Great Barrier Reef," *Newsweek*, https://www .newsweek.com/surfing-robots-are-now-patrolling-great-barrier-reef-683431.

109 **according to Latham's climate models:** John Latham et al., "Marine Cloud Brightening," *The Royal Society* 370, no. 1974 (September 13, 2012), https://doi.org/10.1098 /rsta.2012.0086.

109 **for the journal *Nature* titled "Control of Global Warming?":** John Latham, "Control of Global Warming?," *Nature* 347 (1990): 339–40, https://doi.org/10.1038/347339b0.

109 **markedly brighter, as satellite images prove:** John Latham et al., "Marine Cloud Brightening."

109 **reflect about 10 percent less sunlight:** Goddard Institute for Space Studies, "Cloud Climatology: Global Distribution and Character of Clouds," NASA, https://www .giss.nasa.gov/research/briefs/rossow_01/distrib.html.

109 **bigger-sized water droplets than continental clouds:** Ulrike Lohmann and Glen Lesins, "Comparing Continental and Oceanic Cloud Susceptibilities to Aerosols," *Atmostpheric Science* 30, no. 15 (2003), https://agupubs.onlinelibrary.wiley.com/doi/full/10.1029/2003GL017828.

109 **facilitated by aerosols or pollution:** S. Twomey, "The Influence of Pollution on the Shortwave Albedo of Clouds," *Journal of Atmospheric Science* 34, no. 7 (1977): 1149–52, doi:10.1175/15200469(1977)034<1149:TIOPOT>2.0.CO;2.

110 **"predicted to occur in coming decades," Latham wrote:** John Latham et al., "Marine Cloud Brightening: Regional Applications," *The Royal Society* 371, no. 2031 (2014), https://doi.org/10.1098/rsta.2014.0053.

112 **power the entire global population's energy needs for a year:** "Incoming Sunlight," NASA Earth Observatory, https://earthobservatory.nasa.gov/features/Energy Balance/page2.php.

112 **before making its way to the Earth's surface:** Columbia University, "The Climate System: Solar Radiation, Earth's Atmosphere, and the Greenhouse Effect," Department of Earth and Environmental Sciences, https://eesc.columbia.edu/courses/ees/climate/lectures/radiation_hays.

113 **affects weather a continent or an ocean away:** "Introduction," National Weather Service Climate Prediction Center, https://www.cpc.ncep.noaa.gov/data/teledoc/teleintro.shtml.

113 **changes in the atmospheric wave and jet streams:** "Frequently Asked Questions About El Niño and La Niña," National Weather Service Climate Prediction Center, https://www.cpc.ncep.noaa.gov/products/analysis_monitoring/ensostuff/ensofaq.shtml#NINO.

113 **warmer waters around Christmastime:** "What Are El Niño and La Niña?," NOAA National Ocean Service, https://oceanservice.noaa.gov/facts/ninonina.html.

114 **in the January 2018 journal *Nature Ecology & Evolution*:** Christopher Trisos et al., "Potentially Dangerous Consequences for Biodiversity of Solar Geoengineering Implementation and Termination," *Nature Ecology & Evolution* 2 (2018): 475–82, https://doi.org/10.1038/s41559-017-0431-0.

115 **best chance of saving the Great Barrier Reef:** Jonathan Pearlman, "How Coral of Great Barrier Reef Could Be Saved by Big, Bright Clouds," *Telegraph*, https://www.telegraph.co.uk/news/2017/04/25/coral-great-barrier-reef-could-saved-big-bright-clouds/.

117 **To geoengineer massive seaweed farms around the world:** Tim Flannery, *Sunlight and Seaweed: An Argument for How to Feed, Power and Clean Up the World* (Melbourne: Text Publishing, 2017).

117 **cataclysmic climate events will unfold over the coming century:** Tim Flannery, *The Weather Makers: How Man Is Changing the Climate and What it Means for Life on Earth* (New York: Atlantic Monthly Press, 2005).

118 **the foundation for much of our marine ecosystems:** University of California, Santa Barbara, "Ecology of Seaweed and Its Environmental Significance," Cheadle Center for Biodiversity and Ecological Restoration, https://www.ccber.ucsb.edu/collections-botanical-collections-algae/ecology-seaweed-and-its-environmental-significance.

118 **which reads like a manifesto:** Flannery, *Sunlight and Seaweed*, chap. 8.

120 **and how light scatters:** "SCoPEx," Harvard University Keutsch Research Group, https://projects.iq.harvard.edu/keutschgroup/scopex.

121 **protects us from too much ultraviolet light:** Neville Fletcher, "Earth's Sunscreen, the Ozone Layer," Australian Academy of Science, https://www.science.org.au /curious/earth-environment/earths-sunscreen-ozone-layer.

* John Latham and Stephen Salter have been chosen to work with Cambridge University's new Centre for Climate Repair, which will open in late 2019 and be one of the leading geoengineering research centers in the world: Phyllis Stephen, "Edinburgh Engineer Working to Fix the Polar Ice Caps," *The Edinburgh Reporter*, https:// www.theedinburghreporter.co.uk/2019/06/edinburgh-engineer-working-to-fix -the-polar-ice-caps/.

Chapter 8: Fertilizing Oceans

122 **one of the largest ocean dead zones ever measured:** CBS News, "Gulf of Mexico Dead Zone Is 'Largest' Ever Recorded," *CBS This Morning*, https://www.cbsnews .com/news/gulf-of-mexico-largest-dead-zone-ever-measured-fertilizer/.

123 **40 percent of its fish:** "Gulf of Mexico," Environmental Defense Fund, https://www. edf.org/oceans/gulf-mexico.

123 **fluctuate based on fish hauls:** *Texas Fish & Game Magazine*, "Economic Impact of Fishing Huge, Report Says," FishGame.com, https://fishgame.com/2019/01 /economic-impact-of-fishing-huge-report-says/.

123 **closure of more than a third of the fishing waters:** "Gulf of Mexico Oil Spill," US Food and Drug Administration, https://www.fda.gov/food/food-safety-during -emergencies/gulf-mexico-oil-spill.

123 **has been growing for years:** "Gulf of Mexico 'Dead Zone' Is the Largest Ever Measured," NOAA, https://www.noaa.gov/media-release/gulf-of-mexico-dead-zone-is -largest-ever-measured.

123 **is encouraging greenstick fishing in the Gulf:** "Project to Restore Oceanic Fish Now Underway," NOAA Gulf Spill Restoration, https://www.gulfspillrestoration .noaa.gov/2017/03/project-restore-oceanic-fish-now-underway.

123 **can be quickly released:** Sea Grant Louisiana, "Switching Gears in the Gulf of Mexico—Alternatives to Surface Longlines," LSU Ag Center, https://www .seagrantfish.lsu.edu/management/longlines.htm.

125 **So the torch is always kept:** Prateek Kumar, "In Refineries, Why Do They Have a Constantly Burning Flame at the Top of Some Tower? What Purpose Does It Serve?," Quora, https://www.quora.com/In-refineries-why-do-they-have-a-constantly -burning-flame-at-the-top-of-some-tower-What-purpose-does-it-serve.

125 **according to the most recent numbers available:** NOAA Office of Science and Technology, "Monthly Commercial Landing Statistics," NOAA Fisheries, https://www .fisheries.noaa.gov/national/commercial-fishing/commercial-fisheries-landings.

125 **according to the Food and Agriculture Organization:** Food and Agriculture Organization of the United Nations, *The State of World Fisheries and Aquaculture 2016: Contributing to Food Security and Nutrition for All* (Rome: FAO, 2016), http://www .fao.org/3/a-i5555e.pdf.

126 **continued rising into the foreseeable future:** "Climate Change Indicators: Sea Surface Temperature," US Environmental Protection Agency, https://www.epa.gov /climate-indicators/climate-change-indicators-sea-surface-temperature.

126 **continue at the present rate:** Charles Clover, "All Seafood Will Run Out in 2050, Say Scientists," *Telegraph*, https://www.telegraph.co.uk/news/uknews/1533125/All-seafood -will-run-out-in-2050-say-scientists.html.

126 **overtaken the number of wild fish consumed:** Gwynn Guilford, "The Future Is Here: People Are Now Eating More Farmed Fish than Wild-Caught Fish," *Quartz*, https://qz.com/730794/the-future-is-here-people-are-now-eating-more-farmed -fish-than-wild-caught-fish/.

126 **by millions of square miles since the 1950s:** Denise Breitburg et al., "Declining Oxygen in the Global Ocean and Coastal Waters," *Science* 359, no. 6371 (2018), doi: 10.1126/science.aam7240.

126 **can be catastrophic for marine habitats:** Sarah Moffitt et al., "Response of Seafloor Ecosystems to Abrupt Global Climate Change," *Proceedings of the National Academy of Sciences of the United States of America* 112, no. 15 (2015): 4684–89, https:// doi.org/10.1073/pnas.1417130112.

127 **is produced by marine plants:** Kalila Morsink, "With Every Breath You Take, Thank the Ocean," Smithsonian Ocean, https://ocean.si.edu/ocean-life/plankton/every -breath-you-take-thank-ocean.

127 **create a massive carbon storage layer:** "Phytoplankton," Woods Hole Oceanographic Institution, https://www.whoi.edu/know-your-ocean/ocean-topics/ocean -life/phytoplankton/.

127 **according to marine studies:** Lauren Morello, "Phytoplankton Population Drops 40 Percent Since 1950," *Scientific American*, https://www.scientificamerican.com/article /phytoplankton-population/.

127 **are taken out of the food supply chain:** Clare Leschin-Hoar, "Fish Stocks Are Struggling to Rebound. Why Climate Change Is on the Hook," NPR.org, https://www.npr .org/sections/thesalt/2015/12/14/459404745/fish-stocks-are-declining-worldwide -and-climate-change-is-on-the-hook.

127 **helping phytoplankton to grow there:** Deanna Connors, "Iron from Sahara Dust Helps Fertilize Atlantic Ocean," EarthSky, https://earthsky.org/earth/iron-from -the-sahara-helps-fertilize-atlantic-ocean.

128 **on the surface of the Southern Ocean:** Wendy Zukerman, "Whale Poop Is Vital to Ocean's Carbon Cycle," *New Scientist*, https://www.newscientist.com/article /dn18807-whale-poop-is-vital-to-oceans-carbon-cycle/.

128 **that we already discussed:** A. Paulmier and D. Ruiz-Pino, "Oxygen Minimum Zones (Omzs) in the Modern Ocean," *Progress in Oceanography* 80 (2009): 113–28, doi:10.1016/j.pocean.2008.08.001.

128 **increased in proportion to the amount of iron added:** John Martin, "Iron Deficiency Limits Phytoplankton Growth in the North-East Pacific Subarctic," *Nature* 331 (1988): 341–43, https://doi.org/10.1038/331341a0.

129 **found phytoplankton levels zoomed:** "John Martin," NASA Earth Observatory, https://earthobservatory.nasa.gov/features/Martin/martin_5.php.

129 **can make people sick if they eat shellfish:** Quirin Schiermeier, "The Oresman," *Nature* 421 (2003), https://www.nature.com/articles/421109a.

130 **to abide by government or treaty protocols:** Hugh Powell, "Fertilizing the Ocean with Iron," Woods Hole Oceanographic Institution *Oceanus Magazine*, https:// www.whoi.edu/oceanus/feature/fertilizing-the-ocean-with-iron/.

130 **confirmed a record was set that season:** Cora Campbell, "2013 Salmon Harvest Sets New Records," Alaska Department of Fish and Game, http://www.adfg.alaska.gov /index.cfm?adfg=pressreleases.pr10102013.

131 **false and misleading claims about his research:** "B.C. Company Responds to Ocean Fertilization Lawsuit," CBC News, https://www.cbc.ca/news/canada/british -columbia/b-c-company-responds-to-ocean-fertilization-lawsuit-1.2552119.

132 **in an article called 'The Oresman,'" he says:** Schiermeier, "The Oresman."

132 **You can order it from Amazon:** Red Iron Oxide, 50 lb., Amazon, https://www .amazon.com/Red-Iron-Oxide-50Lb-Natural/dp/B01MDLNBQ6.

133 **the experiment was begun:** Ron Johnson, "Ocean Fertilization Could Be a Boon to Fish Stocks," *Earth Island Journal,* http://www.earthisland.org/journal/index.php /articles/entry/ocean_fertilization_could_be_a_boon_to_fish_stocks/.

133 **famous for his whale-hunting interventions:** Mark Hume, "Sea Shepherd's Watson Vows to Stop B.C. Ocean Fertilization Plan," *Globe and Mail,* https://www .theglobeandmail.com/news/british-columbia/sea-shepherds-watson-vows -to-stop-bc-ocean-fertilization-plan/article5061216/.

135 **to fully restore the sea within fifteen years:** Baltic Sea Restoration, http://www .balticsearestoration.org.

135 **such as the Atlantic Ocean, for dispersion:** "The Baltic Sea—A Unique Marine Region," The Baltic Sea 2020, http://www.balticsea2020.org/english/the-baltic-seas -challanges.

136 **in order to disperse phosphorus:** Lucy Ngatia and Robert Taylor, "Phosphorus Eutrophication and Mitigation Strategies," *Intechopen,* https://www.intechopen.com /online-first/phosphorus-eutrophication-and-mitigation-strategies.

136 **and produces nitrous oxide:** "Stanford Researcher Turns Wastewater into Energy that Can Power Treatment Plants," *KPIX 5,* CBS SF BayArea, https://sanfrancisco .cbslocal.com/2014/03/10/stanford-researcher-turns-wastewater-into-energy-that -can-power-treatment-plants/.

137 **estimated one hundred thousand square miles:** "Dead Zones," William & Mary Virginia Institute of Marine Sciences, https://www.vims.edu/research/topics/dead _zones/index.php.

Chapter 9: Dutch Sea Level Defense

138 **including blizzards, tornadoes, and typhoons:** Christina Nunez, "Floods, Explained," *National Geographic,* https://www.nationalgeographic.com/environment /natural-disasters/floods/.

138 **and put at risk major population centers:** Jessica Shugart, "Flood Damage to Cost Up to $1 Trillion per Year by 2050," *Science News,* https://www.sciencenews.org /article/flood-damage-cost-1-trillion-year-2050.

138 **many of the city's nearly fifteen million residents:** Tran Viet Duc, "Which Coastal Cities Are at Highest Risk of Damaging Floods? New Study Crunches the Numbers," World Bank Group, http://www.worldbank.org/en/news/feature/2013/08/19/coastal -cities-at-highest-risk-floods.

138 **the high mountains of the Himalayas—most at risk:** Dinsa Sachan, "Kolkata

Among World's Most Flood-Prone Coastal Cities: Study," Down To Earth, https://www.downtoearth.org.in/news/kolkata-among-worlds-most-floodprone-coastal-cities-study-38959.

138 **it is so low-lying and has weak infrastructure:** Benjamin Sawe, "Countries Most Prone to Urban Flooding," World Atlas, https://www.worldatlas.com/articles/countries-most-prone-to-urban-flooding.html.

139 **that they are most exposed:** Maldives Tourism, https://maldivestourism.net/maldives/.

139 **an underwater cabinet meeting to highlight its fears:** Mail Foreign Service, "Maldives Government Highlights the Impact of Climate Change . . . by Meeting Underwater," *Daily Mail*, https://www.dailymail.co.uk/news/article-1221021/Maldives-underwater-cabinet-meeting-held-highlight-impact-climate-change.html.

139 **live in critical danger of flooding:** Climate Central, "20 Countries Most at Risk from Sea Level Rise," The Weather Channel, https://weather.com/science/environment/news/20-countries-most-risk-sea-level-rise-20140924.

139 **suffers from widespread inclination, or sinking:** European Space Agency, "Mapping That Sinking Feeling," Phys.org, https://phys.org/news/2016-06-mapping-that-sinking-feeling.html.

139 **indistinguishable bodies of water:** Matt Rosenberg, "How the Netherlands Reclaimed Land from the Sea," ThoughtCo., https://www.thoughtco.com/polders-and-dikes-of-the-netherlands-1435535.

139 **the Westfriese Omringdijk, seventy-five miles long:** "Our History," Dutch Water Sector, https://www.dutchwatersector.com/our-history/.

139 **water management began in earnest:** Ibid.

140 **one of the seven wonders of the modern world:** Ibid.

140 **over the coming decades has now doubled:** Alex Kirby, "Modest Warming Will Raise Europe's Flood Risk," Climate News Network, https://climatenewsnetwork.net/23760-2/.

140 **an eight-foot rise by 2100:** Nicholas Kusnetz, "Sea Level Rise Estimate Grows Alarmingly Higher in Latest Federal Report," Inside Climate News, https://insideclimatenews.org/news/24012017/sea-level-rise-global-warming-federal-report-donald-trump.

140 **to fortify its water defenses:** "What Is the Delta Programme?," Delta Programme Commisioner,https://english.deltacommissaris.nl/delta-programme/what-is-the-delta-programme.

141 **Twelve hundred people died:** Steve George, "A Third of Bangladesh Under Water and Flood Devastation Widens," CNN.com, https://edition.cnn.com/2017/09/01/asia/bangladesh-south-asia-floods/index.html.

141 **Eighty-two people were killed:** "NHS Estimates Harvey Damage at $125 Billion," Occupational Health & Safety, https://ohsonline.com/articles/2018/01/29/nhc-estimates-harvey-damage.aspx; David Fahrenthold, "Residents Warned to 'Get Out or Die' as Harvey Unleashes Waves of Punishing Rains and Flooding," *Washington Post*, https://www.washingtonpost.com/news/post-nation/wp/2017/08/30/harvey-again-makes-landfall-this-time-as-a-tropical-storm-near-cameron-la/?noredirect=on&utm_term=.112b13e45470.

141 **pummeled Argentina and Chile:** Matti Kukkola, "Record Rain, Floods Hit Argentina and Chile," The Watchers, https://watchers.news/2018/04/07/record-rain-floods-hit-argentina-and-chile/.

141 **forcing 175,000 people from their homes:** Humanitarian Aid, "Almost 500,000 Affected as Devastating Floods Inundate Central Somalia—UN mission," UN News, https://news.un.org/en/story/2018/05/1008612.

142 **could more than double that estimate:** Lorenzo Alfieri et al., "Multi-Model Projections of River Flood Risk in Europe Under Global Warming," Climate 6, no. 1 (2018), https://doi.org/10.3390/cli6010006.

142 **costs have mounted into the trillions:** Martin Williams, "2.3 Billion People Affected by Flooding Disasters in 20 Years," Channel 4 News "Fact Check," https://www.channel4.com/news/factcheck/2-3-billion-people-affected-by-flooding-disasters-in-20-years.

142 **chance of such a flood happening during the course of a year:** Robert Holmes, "Floods and Recurrence Intervals," US Geological Survey, https://www.usgs.gov/special-topic/water-science-school/science/floods-and-recurrence-intervals?qt-science_center_objects=0#qt-science_center_objects.

143 **The flood reconfigured much of the state:** Mark Landis, "The Great Flood of 1862 Left Devastation in Its Path Across the State," Sun (San Bernardino, CA), https://www.sbsun.com/2018/11/12/the-great-flood-of-1862-left-devastation-in-its-path-across-the-state/.

143 **linked to the Big One:** "ARkStorm Scenario," US Geological Survey, https://www.usgs.gov/natural-hazards/science-application-risk-reduction/science/arkstorm-scenario?qt-science_center_objects=0#qt-science_center_objects.

144 **and central Europe completely drowned:** Jacqueline Howard, "500-Year Floods Could Strike NYC Every Five Years, Climate Study Says," CNN.com, https://edition.cnn.com/2017/10/27/us/climate-change-new-york-city-floods-study/index.html; Adam Vaughan, "Paris Floods Made Twice as Likely by Climate Change, Say Scientists," Guardian, https://www.theguardian.com/environment/2016/jun/10/paris-floods-made-almost-twice-as-likely-by-climate-change-say-scientists.

144 **whereas other floods can gestate over days:** "Severe Weather 101—Floods," NOAA The National Severe Storms Laboratory, https://www.nssl.noaa.gov/education/svrwx101/floods/types/.

144 **haven't seen rain for million's of years:** Fraser Cain, "What Is the Driest Place on Earth?," Universe Today, https://www.universetoday.com/15031/driest-place-on-earth/.

145 **necessitating bridges to get from one point to another:** "Netherlands Geography," World Atlas, https://www.worldatlas.com/webimage/countrys/europe/netherlands/nlland.htm.

145 **population live in coastal areas:** United Nations, "Factsheet: People and Oceans," in The Ocean Conference: United Nations New York, June 5–7, 2017, https://www.un.org/sustainabledevelopment/wp-content/uploads/2017/05/Ocean-fact-sheet-package.pdf.

146 **residents, visitors, transportation needs, and recreational offerings:** Zones Urbaines Sensibles, https://www.zus.cc/.

146 **a floating city called The Tide City:** Kristian Koreman, "Mission Statement," Presentation from MIT Climate Change Symposium: Sustaining Coastal Cities,

(Cambridge, MA, June 16–18, 2014), https://seagrant.mit.edu/conferences/CCS2014/presentations/235/Kristian_Koreman.pdf.

146 **exists only in model stage:** Elma van Boxel and Kristian Koreman, *Re-public: Towards New Spatial Politics* (Rotterdam: NAi Uitgevers, 2007).

149 **was built from reclaimed land:** Lim Tin Seng, "Land from Sand: Singapore's Reclamation Story," *Biblioasia* 13, no. 1 (April 4, 2017), http://www.nlb.gov.sg/biblioasia/2017/04/04/land-from-sand-singapores-reclamation-story/#sthash.8lDxncMy.dpbs.

149 **the tiny seven islands of Bombay:** "The 7 Islands of Mumbai," Housing.com, https://housing.com/news/the-seven-islands-of-bombay/.

149 **an old, sunken ship buried underneath the city:** Greg Miller, "New Map Reveals Ships Buried Below San Francisco," *National Geographic*, https://news.nationalgeographic.com/2017/05/map-ships-buried-san-francisco/.

149 **cities such as Venice, Italy, and St. Petersburg, Russia:** Soo Kim, "Dubai Unveils Plans to Build Floating Replica of Venice," *Telegraph*, https://www.telegraph.co.uk/travel/destinations/middle-east/united-arab-emirates/articles/dubai-floating-venice-underwater-luxury-resort/.

149 **the toxins can in turn be consumed by fish:** Oliver Milman, "The Facts About Dredging," *Guardian*, https://www.theguardian.com/environment/2013/dec/11/the-facts-about-dredging.

150 **ill effects in sea and on land:** Helena Stratford, "Fish and Heavy Metal Contamination," Pollution Issues, http://www.pollutionissues.co.uk/fish-heavy-metal-contamination.html.

150 **circumnavigate the planet a thousand times:** 22 billion tons equals about a billion dump trucks, which stacked as a caravan reach about 2.5 million miles. The Earth is 25,000 miles in circumference, equating the length of 1,000 trucks.

151 **"likely to amount to several hundred million [people]":** Nobuo Mimura, "Sea-Level Rise Caused by Climate Change and Its Implications for Society," *Proceedings of the Japan Academy, Series B Physical & Biological Sciences* 89, no. 7 (2013): 281–301, doi: 10.2183/pjab.89.281.

152 **inevitable flooding consequences that climate change will bring:** "Vision," Waterstudio.NL, https://www.waterstudio.nl/the-floating-vision-by-koen-olthuis/.

152 **built exclusively for flora and fauna:** "Sea Tree," Waterstudio.NL "Projects," https://www.waterstudio.nl/projects/sea-tree/.

152 **to upgrade the living conditions for waterfront slums:** Ed Hill, "Floating City Apps—Floating Facilities for Flood Prone Areas," *Floodlist*, appearing in Waterstudio.NL "Media," https://www.waterstudio.nl/floating-city-apps-floating-facilities-for-flood-prone-areas/.

154 **are all built into the layers below:** "City in the Sky Concept," Arch20.com, https://www.arch2o.com/city-in-the-sky-concept-tsvetan-toshkov/

154 **populations of more than ten million people:** "City in the Sky," Formad, https://www.formad.co.uk/city-in-the-sky.

154 **the Burj Khalifa in Dubai:** Chris Mills, "Burj Khalifa, Jeddah Tower and the Tallest Buildings of the Future," Owlcation, https://owlcation.com/humanities/The-Quest-for-the-Tallest-Building-in-the-world.

154 **as a solution to a more sustainable future":** Vertical City, https://verticalcity.org/index.html.

154 **dozens more under construction:** "Cities Ranked by Number of 150m+ Completed Buildings," The Skyscraper Center, http://www.skyscrapercenter.com/cities?list= buildings-150.

Chapter 10: Living Beneath the Surface

155 **one of the fastest growing populations in the world:** "Mexico City Population 2019," World Population Review, http://worldpopulationreview.com/world-cities /mexico-city-population/.

155 **is the most populated metro area in the world:** "World City Populations 2019," World Population Review, http://worldpopulationreview.com/world-cities/.

155–156 **on track to add millions more people by mid-century:** John Vidal, "The 100 Million City: Is 21st Century Urbanisation Out of Control?," *Guardian*, https://www .theguardian.com/cities/2018/mar/19/urban-explosion-kinshasa-el-alto-growth -mexico-city-bangalore-lagos.

156 **become the world's fifth-largest economy by then:** "Mexico—the 5th Largest Economy by 2050," Intertraffic.com, https://www.intertraffic.com/news/article/mexico -5th-economy-in-the-world-by-2050/.

156 **five stories in the city's core:** Elizabeth Malkin, "Zoning in Mexico City," Green Economics, http://greeneconomics.blogspot.com/2007/09/zoning-in-mexico -city.html.

157 **often more than 100°F for weeks on end:** "Coober Pedy," Farm Online Weather, http://www.farmonlineweather.com.au/climate/station.jsp?lt=site&lc=16007.

157 **no matter where you are on Earth:** US Department of the Interior, *Geothermal Data of the United States, Including Many Original Determinations of Underground Temperature*, Bulletin 701 (Darton, NH: Washington Government Printing Office, 1920), https://pubs.usgs.gov/bul/0701/report.pdf.

157 **under the surface lies an active volcano:** Christopher Burt, "Warmest Places on Earth: Average Annual Temperature," Weather Underground, https://www .wunderground.com/blog/weatherhistorian/warmest-places-on-earth-average -annual-temperature.html.

157 **may become uninhabitable:** Louise Gray, "Climate Change Could Make Half the World Uninhabitable," *Telegraph*, https://www.telegraph.co.uk/news/earth/envi ronment/climatechange/7710229/Climate-change-could-make-half-the-world -uninhabitable.html.

157 **if our skin temperature exceeds 95°F in wet bulb terms:** Steven Sherwood and Matthew Huber, "An Adaptability Limit to Climate Change Due to Heat Stress," *Proceedings of the National Academy of Sciences of the United States of America* 107, no. 21 (2010): 9552–55, https://doi.org/10.1073/pnas.0913352107.

158 **and heavily populated parts of China:** Ibid.

158 **connected by subway trains and subterranean roads:** The Futurist, "Underground Cities: Japan's Answer to Overcrowding," Questia, https://www.questia.com /magazine/1G1-9177469/underground-cities-japan-s-answer-to-overcrowding.

158 **and an industrial center:** Jess Zimmerman, "Check Out Helsinki's Underground Shadow City," Grist, https://grist.org/urbanism/2011-02-23-check-out-helsinkis -underground-shadow-city/.

158 **been taken over by people in Beijing:** Kaushik, "Beijing's Underground City," Amusing Planet, https://www.amusingplanet.com/2018/11/beijings-underground-city.html.

158 **where more than four thousand people will live:** Calvin Yang, "Singapore Looks Below for More Room," *New York Times,* https://www.nytimes.com/2013/09/26/business/international/crowded-singapore-looks-below-for-room-to-grow.html.

158 **and other commercial spots:** "PATH—Toronto's Downtown Pedestrian Walkway," Toronto.ca, https://www.toronto.ca/explore-enjoy/visitor-services/path-torontos-downtown-pedestrian-walkway/.

158 **underground layers of living and working possibilities:** Jake Nevins, "The Upside Down: Inside Manhattan's Lowline Subterranean Park," *Guardian,* https://www.theguardian.com/cities/2019/apr/06/lowline-park-underground-space-manhattan-new-york-america.

159 **warning of the fallibilities of modern infrastructure:** "Superstudio 50," MAXXI, https://www.maxxi.art/en/events/superstudio-50/.

159 **was to divert rivers:** Stanford University, "The River Arno Project," Stanford.edu, http://leonardodavinci.stanford.edu/projects/arno/Leonardosplan.html.

160 **according to the National Academy of Sciences:** Karen Seto et al., "Global Forecasts of Urban Expansion to 2030 and Direct Impacts on Biodiversity and Carbon Pools," *Proceedings of the National Academy of Sciences of the United States of America* 109, no. 40 (2012): 16083–88, https://doi.org/10.1073/pnas.1211658109.

160 **followed by urban areas in South America:** Nate Berg, "Where Urban Land Growth Is About to Explode," City Lab, https://www.citylab.com/equity/2012/09/where-urban-land-growth-about-explode/3318/.

160 **Delhi; Shanghai; São Paulo; and Mexico City:** "World City Populations 2019," World Population Review.

160 **is expected to more than double to 57 million:** Jeff Desjardins, "Animation: World's Largest Megacities by 2100," Visual Capitalist, https://www.visualcapitalist.com/worlds-20-largest-megacities-2100/.

160 **from managing traffic to transporting food and water efficiently:** "Microsoft CityNext," Microsoft, https://www.microsoft.com/en-us/enterprise/citynext.

160 **through its Sidewalk Labs division:** Sidewalk Labs, https://www.sidewalklabs.com/.

161 **twenty thousand people lived there in the fifth century BC:** "The Lost City of Petra," UN Museum, http://www.unmuseum.org/petra.htm.

161 **lists it as a top threat:** Alex Ward, "The Pentagon Calls Climate Change a National Security Threat. Trump Isn't Listening," Vox, https://www.vox.com/2019/1/18/18188153/pentagon-climate-change-military-trump-inhofe.

162 **the design got picked up in the media:** George Webster, "Could 'Earthscraper' Really Turn Architecture on Its Head?," CNN.com, https://edition.cnn.com/2011/10/27/tech/innovation/earthscraper-mexico-fantasy-reality/index.html.

162 **a seven-story deep underground shopping mall:** Ariel Schwartz, "This 7-Story-Deep Underground Mall Is the Future of Retail," *Business Insider,* https://www.businessinsider.com/mexico-citys-underground-mall-2016-1?r=US&IR=T.

164 **more ancient pyramids in Mexico and the Americas than in all the world:** "Pyramids in Latin America," History.com, https://www.history.com/topics/ancient-history/pyramids-in-latin-america.

164 **about five hundred million people, are severely claustrophobic:** Graham Davey, *Phobias: A Handbook of Theory, Research and Treatment* (New York: Wiley, 1997).

165 **as many as two million people live below ground:** Kieran Nash, "Will We Ever . . . Live in Underground Homes?," BBC.com, http://www.bbc.com/future/story /20150421-will-we-ever-live-underground.

165 **including shopping and entertainment complexes:** Mandy Zuo, "In China, Xi Jinping's New Mega City Xiongan Is Expanding Underground," *South China Morning Post* (Hong Kong), https://www.scmp.com/news/china/policies-politics/article /2099987/parts-xi-jinpings-dream-city-be-built-underground.

166 **under a 400-foot-tall mountain in Rothenstein, Germany:** Vivos, https://terravi vos.com.

167 **to be completely offshore and automated:** "Ocean Spiral: Deep Sea Future City Concept," Shimizu Corporation, https://www.shimz.co.jp/en/topics/dream/content01/; Katharine Tobal, "Japan Releases Plans for Futuristic Underwater Cities by 2030," Collective Evolution, https://www.collective-evolution.com/2014/11/25/plans-for-future -underwater-cities-in-japan-by-2030/.

169 **they built their homes underground:** Jennifer Nalewicki, "Half of the Inhabitants of this Australian Opal Capital Live Underground," *Smithsonian*, https://www .smithsonianmag.com/travel/unearthing-coober-pedy-australias-hidden-city -180958162/.

169 **because they are cheaper and easier to source:** "Coober Pedy: A Model for the Future of Power Generation," Energy Source & Distribution "Latest News," http:// www.esdnews.com.au/coober-pedy-a-model-for-the-future-of-power-generation/.

169 **it's going to be uninhabitable by 2100:** Elizabeth MacBride, "The Hottest Spot on Earth Has a Melting Economy," CNBC.com, https://www.cnbc.com/2016/10/18 /kuwait-is-the-hottest-spot-on-earth.html.

169 **more complex and causes more pollution:** Ruth Michaelson, "Kuwait's Inferno: How Will the World's Hottest City Survive Climate Change?," *Guardian*, https:// www.theguardian.com/cities/2017/aug/18/kuwait-city-hottest-place-earth-climate -change-gulf-oil-temperatures.

170 **"increase by 800 percent to reach 1.6 billion by mid-century":** C40 Cities, *For Cities, the Heat Is On* (C40 Cities, 2017), https://www.c40.org/other/the-future-we-don -t-want-for-cities-the-heat-is-on.

Chapter 11: Stopping the Glaciers from Melting

171 **before freezing yet again come the next winter:** "Arctic vs. Antarctic," National Snow & Ice Data Center, https://nsidc.org/cryosphere/seaice/characteristics/differ ence.html.

171 **decade-over-decade decline in Arctic ice extent:** "Arctic Sea Ice Minimum," NASA Global Climate Change, https://climate.nasa.gov/vital-signs/arctic-sea-ice/.

171 **maybe even a lot sooner:** James Overland and Muyin Wang, "When Will the Summer Arctic Be Nearly Sea Ice Free?," *Geophysical Research Letters* 40, no. 10 (2013): 2097–101, https://doi.org/10.1002/grl.50316.

172 **that connects the Atlantic and Pacific Oceans:** Seth Borenstein, "Arctic Study Discovers that North Pole Not Always Cold," Chron, https://www.chron.com/news

/nation-world/article/Arctic-study-discovers-that-North-Pole-not-always-1526066 .php; Hobart M. King, "What Is the Northwest Passage?," Geology.com, https:// geology.com/articles/northwest-passage.shtml.

172 **cause global sea levels to rise twenty-four feet:** Henry Fountain and Derek Watkins, "As Greenland Melts, Where's the Water Going?," *New York Times*, https:// www.nytimes.com/interactive/2017/12/05/climate/greenland-ice-melting.html.

172 **twice as quickly as the rest of the planet:** "Arctic Report Card," NOAA Arctic Program, https://arctic.noaa.gov/Report-Card.

172 **one hundred feet of coastal erosion:** "Glaciers and Ice Caps to Dominate Sea Level Rise Through 21st Century," National Science Foundation, https://www.nsf.gov /news/news_summ.jsp?cntn_id=109759.

172 **who live along coastal waterways:** Ocean Portal Team, "Sea Level Rise," Smithsonian Ocean, https://ocean.si.edu/through-time/ancient-seas/sea-level-rise.

172 **various sea level scenarios: encroachment, flooding, or total displacement:** Whitney Leach, "Which Cities Are at Risk from Ice Sheets Melting? This Tool Holds the Answers," World Economic Forum, https://www.weforum.org/agenda/2018/06 /this-nasa-tool-can-tell-you-which-melting-glacier-may-flood-your-city/.

173 **with Greenland's northwestern ice areas:** Leanna Garfield, "A New NASA Tool Predicts How High Seas Will Rise in Your City If Specific Glaciers Melt," *Business Insider*, https://www.businessinsider.com/nasa-tool-melting-glaciers-cities-flooding -2017-11?r=US&IR=T.

173 **"the land tends to rise as the ice melts":** John Church et al., "Sea Level Change," in *Climate Change 2013: The Physical Science Basis. Contribution of Working Group I to the Fifth Assessment Report of the Intergovernmental Panel on Climate Change* (Cambridge, UK: Cambridge University Press, 2013), https://www.ipcc.ch/site/assets/up loads/2018/02/WG1AR5_Chapter13_FINAL.pdf.

173 **one thousand feet in certain places:** "Land Rise and Sea Level," Swedish National Knowledge Centre for Climate Change Adaptation, http://www.smhi.se/en/theme /land-rise-and-sea-level-1.12265.

174 **causing more than two hundred feet of sea level rise:** "Quick Facts on Ice Sheets," National Snow & Ice Data Center, https://nsidc.org/cryosphere/quickfacts/icesheets .html.

174 **six feet six inches by the end of the century:** "Understanding Sea Level," NASA Sea Level Change, https://sealevel.nasa.gov/understanding-sea-level/overview.

174 **would affect 145 million people:** Anjana Ahuna, "145 Million Live Within Three Feet of Sea Level. Rising Oceans Are a First World Problem Too," *New Statesman*, https://www.newstatesman.com/culture/books/2018/03/145-million-live-within -three-feet-sea-level-rising-oceans-are-first-world.

174 **a third of southern Florida will vanish:** Terrell Johnson, "As Sea Levels Rise, Is Miami Doomed?," The Weather Channel, https://weather.com/science/environment /news/sea-level-rises-miami-doomed-20130625.

174 **will be at major risk of flooding:** Warren Cornwall, "As Sea Levels Rise, Bangladesh Islanders Must Decide Between Keeping the Water Out—Or Letting It In," *Science*, https://www.sciencemag.org/news/2018/03/sea-levels-rise-bangladeshi-islanders -must-decide-between-keeping-water-out-or-letting.

174 **sitting neatly within the Arctic Circle:** "Experience Svartisen," Svartisen Glacier, http://www.svartisen.no/?page_id=130&lang=en.

175 **reach over six thousand feet high:** "Svartisen Glacier," GoNorway, http://www.gonorway.no/norway/articles/87.

175 **of any such ice formation in Europe:** "Glaciers," Visit Norway, https://www.visitnorway.com/things-to-do/nature-attractions/glaciers/.

175 **not seen since the Little Ice Age:** Wilfred Theakstone, "Glacier Changes at Svartisen, Northern Norway, During the Last 125 Years: Influence of Climate and Other Factors," *Journal of Earth Science* 21, no. 2 (2010): 123–36, http://en.earth-science.net/PDF/20180715112759.pdf.

175 **Glaciers the world over are receding:** Seth Borenstein, "Then and Now: How Glaciers Around the World Are Melting," Phys.org, https://phys.org/news/2017-04-glaciers-world.html.

175 **more glaciers melt and melt fast:** Bethan Davies, "Mapping the World's Glaciers," Antarctic Glaciers, http://www.antarcticglaciers.org/glaciers-and-climate/glacier-recession/mapping-worlds-glaciers/.

175 **as well as to power electricity:** Damian Carrington, "A Third of Himalayan Ice Cap Doomed, Finds Report," *Guardian*, https://www.theguardian.com/environment/2019/feb/04/a-third-of-himalayan-ice-cap-doomed-finds-shocking-report.

175 **even the glaciers there are receding:** John Vidal, "Most Glaciers in Mount Everest Area Will Disappear with Climate Change—Study," *Guardian*, https://www.theguardian.com/environment/2015/may/27/most-glaciers-in-mount-everest-area-will-disappear-with-climate-change-study.

176 **is locked up in glaciers and ice caps:** "Ice, Snow, and Glaciers and the Water Cycle," US Geographical Survey, https://www.usgs.gov/special-topic/water-science-school/science/ice-snow-and-glaciers-and-water-cycle.

176 **forcing them into retreat:** University of Colorado Boulder, "Global Glaciers, Ice Caps, Shedding Billions of Tons of Mass Annually," ScienceDaily, https://www.sciencedaily.com/releases/2012/02/120208132301.htm.

176 **"remain in balance or even grow," the NSIDC explains:** "The Life of a Glacier," National Snow & Ice Data Center, https://nsidc.org/cryosphere/glaciers/life-glacier.html.

176 **Water reflects just 10 percent:** Energy Balance, "Snow Albedo," Climate Policy Watcher, https://www.climate-policy-watcher.org/energy-balance/snow-albedo.html.

177 **causing ice and snow to melt more quickly:** "Black and White: Soot on Ice," NASA, https://www.nasa.gov/vision/earth/environment/arctic_soot.html.

177 **adding 25 percent more greenhouse gas emissions to the atmosphere:** Alessandra Potenza, "Here's What Vanishing Sea Ice in the Arctic Means for You," The Verge, https://www.theverge.com/2018/5/10/17339046/arctic-sea-ice-decline-albedo-effect-climate-change-global-warming; K. Pistone et al., "Observational Determination of Albedo Decrease Caused by Vanishing Arctic Sea Ice," *Proceedings of the National Academy of Sciences of the United States of America* 111 (2014): 3322–26, doi.org/10.1073/pnas.1318201111.

177 **swamping many of the land areas of the planet:** "Facts About Glaciers," National Snow & Ice Data Center, https://nsidc.org/cryosphere/glaciers/quickfacts.html.

177 **according to analysis published in *Nature Climate Change*:** Adrian Raftery et al., "Less than 2°C Warming by 2100 Unlikely," *Nature Climate Change* 7 (2017): 637–41, https://doi.org/10.1038/nclimate3352.

178 **to stall the fastest flows of ice into the ocean:** John Moore et al., "Geoengineer Polar

Glaciers to Slow Sea-Level Rise," *Nature*, https://www.nature.com/articles/d41586 -018-03036-4.

179 **how to stop ice sheets from quickly disappearing:** Hanne Bakke, "World's Most Claustrophobic Lab," ScienceNordic, http://sciencenordic.com/world%E2%80%99s -most-claustrophobic-lab.

181 **instead of using the long route south around Africa:** David McDonald, "The Northwestern Passage: A Great Economic Opportunity, but not Without Environmental Sacrifice," Foreign Policy News, http://foreignpolicynews.org/2016/12/22 /northwestern-passage-great-economic-opportunity-not-without-environmental -sacrifice/.

182 **the leading source of weather information for the planet:** Michael Kuhne, "How Rapid Arctic Sea Ice Melt May Alter Global Weather Patterns," AccuWeather, https:// www.accuweather.com/en/weather-news/how-rapid-arctic-sea-ice-melt-may-alter -global-weather-patterns/70000514#menu-country.

183 **often used for animal feed as well:** "Our Solutions," Ice911, https://www.ice911.org /beads.

183 **and increase ice thickness:** "Climate Modeling," Ice911, https://www.ice911.org /climate-modeling.

184 **would amount to hundreds of millions of dollars:** Leslie Field, *Soft Geoengineering: Ice911* (Ice911 Research Corporation, 2012), https://www.wilsoncenter.org/sites /default/files/Leslie%20Field%20Ice911%20Soft%20Geoengineering%20Panel %20110712%20main.pdf.

184 **saving it for when water is in higher demand for agriculture:** The Ice Stupa Project, http://icestupa.org/.

186 **the number of artificial glaciers built every year:** Sonam Wangchuk, "We Won the Rolex Award for Enterprise," The Ice Stupa Project, http://icestupa.org/news/we -wrolex-awards-for-enterprise.

Chapter 12: What Lies Beneath

189 **yet it is suffering from thirst:** "Brazil May Be the Owner of 20% of the World's Water Supply but It Is Still Very Thirsty," The World Bank, http://www.worldbank .org/en/news/feature/2016/07/27/how-brazil-managing-water-resources-new -report-scd.

189 **sees heavy rainfall:** Alexei Barrionuevo, "Whose Rain Forest Is This, Anyway?" *New York Times*, https://www.nytimes.com/2008/05/18/weekinreview/18barrionu evo.html.

189 **three-quarters of the total freshwater supply:** "Brazil May Be the Owner of 20% of the World's Water Supply but It Is Still Very Thirsty," The World Bank.

189 **The national debt is huge:** 2019 Index of Economic Freedom, "Brazil," Heritage.org, https://www.heritage.org/index/country/brazil.

190 **infrastructure projects throughout the country:** "Investment in Infrastructure Tops the Policy Agenda in Brazil," Global Infrastructure Hub, https://www.gihub .org/blog/investment-in-infrastructure-tops-the-policy-agenda-in-brazil/.

191 **delayed again and again:** Mark Byrnes, "Amid Brazil's Stadium Boom, a Massive Water Infrastructure Project Drags on," City Lab, https://www.citylab.com/environ-

ment/2014/02/amidst-brazils-stadium-boom-massive-water-infrastructure-drags
/8364/.

191 **expected to be completed soon after:** Jake Spring, "Brazil to Launch Canal to
Drought-Stricken Northeast This Year: Minister," *Business Insider*, https://www
.businessinsider.com/r-brazil-to-launch-canal-to-drought-stricken-northeast-this
-year-minister-2018-3?r=US&IR=T; Special Secretariat of Communication of the
Presidency of the Republic, "Water Safety," Brazil.gov, http://www.brasil.gov.br
/valeubrasil/textos/eixo-social-e-cidadania/seguranca-hidrica.

191 **beginning in 1939:** "Delaware Aqueduct Rondout—West Branch Bypass Tunnel
Project," Water Technology, https://www.water-technology.net/projects/delaware
-aqueduct-rondout-west-branch-bypass-tunnel/.

191 **and is lined with concrete:** Institute of Policy Studies, *Kashmir Watch* (Islamabad,
Pakistan: Kashmir Watch Group, 1994).

191 **at least the past twenty-five years:** Susan Xu, "$30 Million Tunnel-Boring Machine
Will Fix Leaking Delaware Aqueduct that Carries NYC's Water," Untapped Cities,
https://untappedcities.com/2018/01/24/30-million-tunnel-boring-machine-begins
-job-to-fix-leaking-delaware-aqueduct/.

191 **because of poor pipes and plumbing:** "Global Water Loss: 45 Billion Litres of Water
Every Day," iSoil Industria Online, http://isoilonline.com/2018/01/water-loss-45
-billion-litres/.

191 **this loss is fourteen billion dollars a year:** Heather Clancy, "With Annual Losses
Estimated at $14 Billion, It's Time to Get Smarter About Water," *Forbes*, https://www
.forbes.com/sites/heatherclancy/2013/09/19/with-annual-losses-estimated
-at-14-billion-its-time-to-get-smarter-about-water/#1b170a6c98a2.

191 **access at all to fresh water:** "2.1 Billion People Lack Safe Drinking Water at Home,
More than Twice as Many Lack Safe Sanitation," World Health Organization, https://
www.who.int/news-room/detail/12-07-2017-2-1-billion-people-lack-safe-drinking
-water-at-home-more-than-twice-as-many-lack-safe-sanitation.

192 **upgrade and fix its leaky aqueduct:** "NYC DEP Starts $1 Billion Tunneling Project
to Repair Delaware Aqueduct," *Civil + Structural Engineer*, https://csengineermag
.com/nyc-dep-starts-1-billion-tunneling-project-repair-delaware-aqueduct/.

192 **a college degree in civil engineering:** Syl Kacapyr, "NYC Tunnel-Borer Named
for Cornell Engineer, Suffragist," *Cornell Chronicle*, http://www.news.cornell.edu
/stories/2017/03/nyc-tunnel-borer-named-cornell-engineer-suffragist.

192 **a vast network of canals, aqueducts, and reservoirs:** Kempton Webb and Concep-
ción Castañeda, "São Francisco River," Encyclopedia Britannica, https://www
.britannica.com/place/Sao-Francisco-River.

193 **twelve million urban dwellers as well as rural farmers:** Kayla Ritter, "São Paulo
Heading to Another Dry Spell," Circle of Blue, https://www.circleofblue.org/2018
/water-climate/drought/sao-paulo-heading-to-another-dry-spell/.

193 **ponds, lakes, wells, streams, or what have you:** Soofia Mahmood, "How Much Do
People Walk for Water?," Ecoloodi, http://ecoloodi.org/en/people-walk-water/.

193 **in developing countries around the globe:** Vicky Hallett, "Millions of Women Take
a Long Walk with a 40-Pound Water Can," NPR.org, https://www.npr.org/sections
/goatsandsoda/2016/07/07/484793736/millions-of-women-take-a-long-walk
-with-a-40-pound-water-can; "2.1 Billion People Lack Safe Drinking Water at Home,
More than Twice as Many Lack Safe Sanitation," World Health Organization, https://

www.who.int/news-room/detail/12-07-2017-2-1-billion-people-lack-safe-drinking-water-at-home-more-than-twice-as-many-lack-safe-sanitation.

193 **They are primitive:** Diego Tales Da Silva, "Brazil Water for All!," Waves for Water, https://www.wavesforwater.org/courier/brazil-water-for-all.

194 **just south of Tulum:** "World's Largest Underwater Cave Discovered," *National Geographic,* https://news.nationalgeographic.com/2018/01/largest-underwater-cave-system-in-the-world-discovered-in-mexico-spd/.

194 **is drying up, according to geologists:** Char Miller, "Farmers Are Drawing Groundwater from the Giant Ogallala Aquifer Faster than Nature Replaces It," The Conversation, https://theconversation.com/farmers-are-drawing-groundwater-from-the-giant-ogallala-aquifer-faster-than-nature-replaces-it-100735.

195 **never mind the gas and diesel utilized:** Claudia Copeland and Nicole Carter, *CRS Report R43200: Energy-Water Nexus: The Water Sector's Energy Use* (Washington, DC: Congressional Research Service, 2017), https://fas.org/sgp/crs/misc/R43200.pdf.

195 **their only function was to block water:** Heloisa Yang et al., "The History of Dams," University of California, Davis, 1999, https://watershed.ucdavis.edu/shed/lund/dams/Dam_History_Page/History.htm.

195 **masonry dams, overflow dams, afterbay dams, dikes:** "Types of Dams," US Society on Dams, https://www.ussdams.org/dam-levee-education/overview/types-of-dams/.

195 **rise more than four stories tall:** "Questions and Answers About Large Dams," International Rivers, https://www.internationalrivers.org/questions-and-answers-about-large-dams.

195 **which is the longest in Asia:** Traci Pedersen, "Yangtze River: Longest River in Asia," Live Science, https://www.livescience.com/57905-yangtze-river-facts.html.

196 **meets ten percent of China's energy demand:** Brian Handwerk, "China's Three Gorges Dam, by the Numbers," *National Geographic,* https://www.nationalgeographic.com/science/2006/06/china-three-gorges-dam-how-big/.

196 **Dams throw ecosystems out of whack:** "Environmental Impacts," Mount Holyoke College The Three Gorges Dam Project, https://www.mtholyoke.edu/~lpohara/Pol%20116/enviro.html.

196 **stop the construction of more large dams:** "The International Anti-Dam Movement," International Rivers, https://www.internationalrivers.org/the-international-anti-dam-movement.

196 **a blog post on April 9, 2018, for Patagonia's website:** Yvon Chouinard, "Telling the Dam Truth," Patagonia, https://www.patagonia.com/blog/2018/04/telling-the-dam-truth/.

197 **consequences, including massive floods:** Alexis Madrigal, "What We've Done to the Mississippi River: An Explainer," *Atlantic,* https://www.theatlantic.com/technology/archive/2011/05/what-weve-done-to-the-mississippi-river-an-explainer/239058/.

197 **construction began a decade later:** NYC Environmental Protection, "History of New York City's Water Supply System," NYC.gov, https://www1.nyc.gov/html/dep/html/drinking_water/history.shtml.

197 **an unsung marvel of service:** "Delaware Tunnel Origin, New York," The Center for Land Use Interpretation, http://www.clui.org/ludb/site/delaware-tunnel-origin.

197 **as a way to protect sediment flow:** Department of Environmental Conservation, "Ashokan Reservoir," NY.gov, https://www.dec.ny.gov/outdoor/101552.html.

198 **deliver unfiltered drinking water to its residents:** "New York City Approved to Supply Unfiltered Drinking Water," *Water Quality Products*, https://www.wqpmag .com/treatment/new-york-city-approved-supply-unfiltered-drinking-water.

198 **the most sophisticated water-testing devices:** Christopher Lee, "Bluegill on Guard in Region's Water Supply," *Washington Post*, http://www.washingtonpost.com/wp -dyn/content/article/2006/09/17/AR2006091700406.html.

199 **it trickles, streams, and drips out of taps:** NYC Environmental Protection, "New York City's Water Supply System Map," NYC.gov, https://www1.nyc.gov/html/dep /html/drinking_water/wsmaps_wide.shtml—elevations derived from these.

199 **soil reacts differently, and can cave:** Mao-Hong Yu and Jian-Chun Li, "Underground Caves, Tunnels, and Excavation of Hydraulic Power Station," in *Computational Plasticity,* Advanced Topics in Science and Technology in China (Berlin: Springer, 2012), https://link.springer.com/chapter/10.1007%2F978-3-642-24590-9_10.

201 **built near or around bodies of water:** "Why Are Cities Located on or Near Rivers?," Portland State University, https://www.pdx.edu/geography-education/sites/www .pdx.edu.geography-education/files/Why%20Are%20Cities%20Located%20On% 20or%20Near%20Rivers%20%28Katelyn%20Ion%29.pdf.

203 **Taklamakan Desert floor in northwest China:** Sofia Lotto Persio, "China Plans World's Longest Tunnel to Turn Desert into California," *Newsweek*, https://www .newsweek.com/china-plans-worlds-longest-tunnel-turn-desert-california-696326.

203 **a terrific amount:** Stephen Chen, "Chinese Engineers Plan 1,000km Tunnel to Make Xinjiang Desert Bloom," *South China Morning Post* (Hong Kong), https://www.scmp .com/news/china/society/article/2116750/chinese-engineers-plan-1000km-tunnel -make-xinjiang-desert-bloom.

204 **government's ambitious plans for growth:** "Private Participation in Infrastructure Database: Country Snapshot," The World Bank, https://ppi.worldbank.org/ppinew /snapshots/country/china.

204 **the biggest "sand sea" in the world:** Martin Rittner et al., "The Provenance of Taklamakan Desert Sand," *Earth and Planetary Science Letters* 437 (2016): 127–37, https://doi.org/10.1016/j.epsl.2015.12.036.

Chapter 13: Food Animals

205 **on Planet Earth:** "Current World Population," WorldoMeters, https://www.world ometers.info/world-population/.

205 **Eritrea, Djibouti, Kenya, Sudan, and Uganda:** Food and Agriculture Organization of the United Nations, *The State of Food Security and Nutrition in the World 2017: Building Resilience for Peace and Food Security* (Rome: FAO, 2017), https://docs.wfp .org/api/documents/WFP-0000022419/download/?_ga=2.268521401.974489353 .1506032202-1929593996.1506032202.

205 **with a population of 160 million:** Food and Agriculture Organization of the United Nations, "Food Insecurity in the Horn of Africa," in *The Elimination of Food Insecurity in the Horn of Africa: Final Report* (Rome: FAO, September 30, 2000), http:// www.fao.org/3/x8530e/x8530e02.htm.

205 **as it did in South Sudan in 2017:** "South Sudan Declares Famine in Unity State," BBC.com, https://www.bbc.co.uk/news/world-africa-39025927.

205 **as it did in Malawi in 2015:** "2015 Floods Leave Malawi Facing Worst Food Crisis in 10 Years," Floodlist, http://floodlist.com/africa/floods-malawi-facing-worst-food -crisis-10-years.

205 **caused the world's worst food crisis:** "Yemen War Causing World's Worst Food Crisis," Norwegian Refugee Council, https://www.nrc.no/yemen-war-causing-worlds -worst-food-crisis.

205 **under the age of five dies there:** Daniel Nikbakht and Sheena McKenzie, "The Yemen War Is the World's Worst Humanitarian Crisis, UN Says," CNN.com, https:// edition.cnn.com/2018/04/03/middleeast/yemen-worlds-worst-humanitarian -crisis-un-intl/index.html.

205 **the most vulnerable in the world:** Jeffrey Sachs, "The Horn of Africa Crisis Is a Warning to the World," *Guardian*, https://www.theguardian.com/global-development /poverty-matters/2011/jul/28/horn-africa-drought-warning.

206 **rate of child malnourishment:** Xinhua, "Ethiopia's Child Malnutrition Rate Down 20 Percentage Points in 16 Years: UNICEF," New China, http://www.xinhuanet .com/english/2017-07/20/c_136459512.htm.

206 **nearly a third from 2000 to 2018:** "Global Hunger Index 2018: Forced Migration and Hunger," The International Food Policy Research Group, https://www.global hungerindex.org/pdf/en/2018.pdf.

206 **and eradicate world hunger:** "Break Down Barriers to End World Hunger by 2030," Action Against Hunger, https://www.actionagainsthunger.org/blog/break-down -barriers-end-world-hunger-2030.

206 **on less than $1.25 a day:** "MDG1: Eradicate Extreme Poverty and Hunger," MDG Monitor, https://www.mdgmonitor.org/mdg-1-eradicate-poverty-hunger/.

207 **to degrade more quickly:** Amy Schweikert, "Climate Change and Infrastructure Impacts: Comparing the Impact on Roads in 10 Countries Through 2100," *Procedia Engineering* 78 (2014): 306–16, doi.org/10.1016/j.proeng.2014.07.072.

207 **hottest in the world in the deserts:** World Meteorological Organization, *WMO-No. 1147: The Climate in Africa: 2013* (Geneva, Switzerland: WMO, 2015), https://library .wmo.int/doc_num.php?explnum_id=3197.

207 **can melt the soles of your shoes:** Josh Sampiero, "Explore Africa's Most Breathtaking Volcano," Red Bull, https://www.redbull.com/us-en/breathtaking-photos-of-the -dallol-volcano-in-ethiopia.

207 **are acutely food insecure:** Meghan Orner, "Malnutrition in Africa," The Borgen Project, https://borgenproject.org/malnutrition-ethiopia/.

207 **lack adequate food:** Denise Koenig, "5 Worst Spots for Hunger," World Vision, https://www.worldvision.org/hunger-news-stories/5-worst-spots-hunger.

208 **"where one in eight is hungry":** "What Causes Hunger?," World Food Programme, https://www.wfp.org/stories/what-causes-hunger.

208 **along with other disruptions such as harsh weather:** Georgina Gustin, "Climate Change Could Lead to Major Crop Failures in the World's Biggest Corn Regions," Inside Climate News, https://insideclimatenews.org/news/11062018/climate-change -research-food-security-agriculture-impacts-corn-vegetables-crop-prices.

208 **"severe environmental impacts," the FAO says:** "Feeding Nine Billion in 2050," Food and Agriculture Organization of the United Nations, http://www.fao.org /news/story/en/item/174172/icode/.

209 **forty million Americans face hunger every day:** "Hunger in America, the Facts," Feeding America, https://www.feedingamerica.org/hunger-in-america/facts.

209 **to stamp out hunger in America:** The Howard G. Buffett Foundation, http://www.thehowardgbuffettfoundation.org/.

209 **to increase crop yields:** Nina Munk, "How Warren Buffett's Son Would Feed the World," *Atlantic*, https://www.theatlantic.com/magazine/archive/2016/05/how-warren-buffetts-son-would-feed-the-world/476385/.

209 **more than four billion dollars:** Betty Liu, "Howard Buffett Is Getting His Hands Dirty," Bloomberg.com, https://www.bloomberg.com/features/2015-howard-buffett/.

209 **was introduced into the market:** Gabriel Rangel and Anna Maurer, "From Corgis to Corn: A Brief Look at the Long History of GMO Technology," Harvard University, http://sitn.hms.harvard.edu/flash/2015/from-corgis-to-corn-a-brief-look-at-the-long-history-of-gmo-technology/.

210 **common GMOs are soy, corn, and sugar beets:** Chris Keenan, "Top 10 Most Common GMO Foods," The Cornucopia Institute, https://www.cornucopia.org/2013/06/top-10-most-common-gmo-foods/.

210 **not their genetic structure:** J. L. Vicini, "GMO Crops in Animal Nutrition," *Animal Frontiers* 7, no. 2 (2017): 9–14, doi.org/10.2527/af.2017.0113.

210 **into the human food chain:** Andrew Pollack, "FDA Approves Drugs from Gene-Altered Goats," *New York Times*, https://www.nytimes.com/2009/02/07/business/07goatdrug.html.

210 **and brought it for regulatory approval in the USA:** AquaBounty, https://aquabounty.com/.

211 **are either at or near depletion:** "Decreasing Fish Stocks," WWF, http://wwf.panda.org/knowledge_hub/endangered_species/cetaceans/threats/fishstocks/.

211 **into the eggs of three thousand goldfish:** "World's First Transgenic Fish," Institute of Hydrobiology, Chinese Academy of Sciences, http://english.ihb.cas.cn/rh/as/201001/t20100112_49951.html.

212 **as food for bears and other animals:** Genny Anderson, "Salmon Reproduction," Marine Science, http://www.marinebio.net/marinescience/05nekton/sarepro.htm.

212 **return to the same spawning grounds for years:** "Atlantic Salmon," Oceana, https://oceana.org/marine-life/ocean-fishes/atlantic-salmon.

212 **and is sent to the manufacturing facility:** Richard Martin, "How Genetically Engineered Salmon Swims onto Our Plates," GreenBiz, https://www.greenbiz.com/article/how-genetically-engineered-salmon-swims-our-plates.

213 **also has facilities in Panama and Prince Edward Island:** Matthew Gonzales, "The World's First GMO Fish Is Stranded in Albany, Indiana," *Indianapolis Monthly*, https://www.indianapolismonthly.com/news-and-opinion/news/the-worlds-first-gmo-fish-is-stranded-in-albany-indiana-2.

213 **Senator Murkowski said in a prepared statement:** Lisa Murkowski, "Murkowski Statement on New US Genetically Engineered Salmon Facility," Murkowski. Senate.gov, https://www.murkowski.senate.gov/press/release/murkowski-statement-on-new-us-genetically-engineered-salmon-facility-.

214 **to learn if it is genetically modified:** Dan Charles, "Congress Just Passed a GMO Labeling Bill. Nobody's Super Happy About it," NPR.org, https://www.npr.org/sections/thesalt/2016/07/14/486060866/congress-just-passed-a-gmo-labeling-bill-nobodys-super-happy-about-it.

214 **10 percent are pro-GMO:** Cary Funk and Brian Kennedy, "The New Food Fights: US Public Divides over Food Science," Pew Research Center, https://www.pewresearch.org/science/2016/12/01/public-opinion-about-genetically-modified-foods-and-trust-in-scientists-connected-with-these-foods/.

214 **include a strict environmental risk assessment:** Commission Directive, "Commission Directive EU 2018/350," *Official Journal of the European Union* 1371 (2018), https://eur-lex.europa.eu/legal-content/EN/TXT/?uri=CELEX:32018L0350.

214 **many refer to GM foods as "frankenfood":** "Frankenfood: Does It Deserve the Name?," CBS News, https://www.cbsnews.com/pictures/frankenfood-does-it-deserve-the-name/.

214 **standards are tightening:** Wendan Wang, "International Regulations on Genetically Modified Organisms: US, Europe, China and Japan," *Food Safety Magazine*, https://www.foodsafetymagazine.com/magazine-archive1/junejuly2016/international-regulations-on-genetically-modified-organisms-us-europe-china-and-japan/.

215 **or to treat genetic causes of disease:** "Questions and Answers About CRISPR," Broad Institute, https://www.broadinstitute.org/what-broad/areas-focus/project-spotlight/questions-and-answers-about-crispr.

216 **"the failed blueprints from *Jurassic Park*":** Tim Schwab, "Science or Science Fiction? Jurassic Park, GMO Salmon and the FDA," Food & Water Watch, https://www.foodandwaterwatch.org/news/science-or-science-fiction-jurassic-park-gmo-salmon-and-fda.

216 **is a "dangerous experiment":** "AquaBounty Fined for Repeated Environmental Violations of Genetically Engineered Salmon," Center for Food Safety, https://www.centerforfoodsafety.org/press-releases/3570/aquabounty-fined-for-repeated-environmental-violations-on-genetically-engineered-salmon.

216 **that will be impossible to turn back:** "FDAs Approval of GMO Salmon Denounced," Friends of the Earth, https://foe.org/news/2015-11-fdas-approval-of-gmo-salmon-denounced/.

216 **largely disagree with the alacrity:** Jane Brody, "Are GMO Foods Safe?" *New York Times*, https://www.nytimes.com/2018/04/23/well/eat/are-gmo-foods-safe.html.

217 **could wipe out the entire species:** Bill Muir, "Transgenic Fish Could Threaten Wild Populations," Purdue University, https://www.purdue.edu/uns/html4ever/0002.Muir.trojan.html.

217 **the famous line stated above:** "Soylent Green," IMDb, https://www.imdb.com/title/tt0070723/.

218 **described as full, ready-to-drink meals:** Soylent, https://soylent.com/.

219 **or other variations can be made from it:** Memphis Meats, http://www.memphismeats.com.

220 **in the range of twenty-four hundred dollars:** Zara Stone, "The High Cost of Lab to Table Meat," *Wired*, https://www.wired.com/story/the-high-cost-of-lab-to-table-meat/.

220 **to increase in the next decade:** Marta Zaraska, "Lab-Grown Meat Is in Your Future, and It May Be Healthier than the Real Stuff," *Washington Post*, https://www.washingtonpost.com/national/health-science/lab-grown-meat-is-in-your-future-and-it-may-be-healthier-than-the-real-stuff/2016/05/02/aa893f34-e630-11e5-a6f3-21ccdbc5f74e_story.html?noredirect=on&utm_term=.1e59b548b9c5.

220 **according to the Food and Agricultural Organization of the United Nations:** "Major Gains in Efficiency of Livestock Systems Needed," Food and Agriculture Organization of the United Nations, http://www.fao.org/news/story/en/item/116937 /icode/.

Chapter 14: Toilet to Tap

221 **most fresh water to people on Earth:** Matti Kummu et al., "How Close Do We Live to Water? A Global Analysis of Population Distance to Freshwater Bodies," *PLoS One* 6, no. 6 (2011), doi: 10.1371/journal.pone.0020578.

221 **kill, mentally and physically impair, and breed birth defects:** Chris Woodford, "Water Pollution: An Introduction," Explain That Stuff, https://www.explainthat stuff.com/waterpollution.html.

221 **as their main water source for drinking and irrigation:** Dikanaya Tarahita and Muhammad Rakhmat, "Indonesia's Citarum: The World's Most Polluted River," *Diplomat*, https://thediplomat.com/2018/04/indonesias-citarum-the-worlds -most-polluted-river/.

221 **For them, it is holy water:** Danish Siddiqui, "Dying 'Mother Ganga,' India's Holy River Succumbs to Pollution," Reuters, https://www.reuters.com/article/us-india -ganges/dying-mother-ganga-indias-holy-river-succumbs-to-pollution -idUSKBN19V0OG; Victor Mallet, "The Ganges: Holy, Deadly River," *FT Magazine*, https://www.ft.com/content/dadfae24-b23e-11e4-b380-00144feab7de.

221 **due to all of its untreated supply of effluent:** T. De Pippo et al., "The Case of the Sarno River (Southern Italy): Effects of Geomorphology on the Environmental Impacts," *Environmental Science and Pollution Research International* 13, no. 3 (2006): 184–91, https://www.ncbi.nlm.nih.gov/pubmed/16758709.

222 **to the toxicity:** "Flint Water Crisis Fast Facts," CNN.com, https://edition.cnn.com /2016/03/04/us/flint-water-crisis-fast-facts/index.html; Cedric Taylor, "Flint's Water Now Safe to Drink—but the Crisis Has Corroded Resident's Trust in Government," City Metric, https://www.citymetric.com/politics/flint-s-water-now-safe -drink-crisis-has-corroded-residents-trust-government-3895.

223 **they end up absorbing more lead instead:** "Lead Poisoning and Health," World Health Organization, https://www.who.int/news-room/fact-sheets/detail/lead -poisoning-and-health.

224 **water use spiked by a factor of fifteen:** Charles Vörösmarty and Dork Sahagian, "Anthropogenic Disturbance of the Terrestrial Water Cycle," *BioScience* 50, no. 9 (2000): 753–65, doi.org/10.1641/0006-3568(2000)050[0753:ADOTTW]2.0.CO;2.

225 **is for our food and agriculture:** "How It Works: Water for Electricity," Union of Concerned Scientists, https://www.ucsusa.org/clean-energy/energy-water-use /water-energy-electricity-overview.

225 **(about 1.6 billion gallons) of water per year:** Yevgeniy Sverdlik, "Here's How Much Water All US Data Centers Consume," Data Knowledge Center, http://www .datacenterknowledge.com/archives/2016/07/12/heres-how-much-water-all-us -data-centers-consume.

225 **water pollution is on the rise:** World Water Assessment Programme, "Facts and

Figures: Water Pollution Is on the Rise Globally," UNESCO, http://www.unesco.org/new/en/natural-sciences/environment/water/wwap/facts-and-figures/all-facts-wwdr3/fact-15-water-pollution/.

225 **on the entire planet for the taking:** Jason Major, "Earth Has Less Water Than You Think," phys.org, https://phys.org/news/2012-05-earth.html.
* The United States Geological Survey created a visual representation of this: "All of Earth's Water in a Single Sphere!," United States Geological Survey, https://www.usgs.gov/media/images/all-earths-water-a-single-sphere.

225 **leaving ground sources with less water:** "Evaporation," Water and Global Change, http://www.waterandclimatechange.eu/evaporation.

225 **for everyone on the planet to use:** Mesfin Mekonnen and Arjen Hoekstra, "Four Billion People Facing Severe Water Scarcity," *Science Advances* 2, no. 2 (2016), doi: 10.1126/sciadv.1500323.

225 **some kind of water scarcity problem:** Jonathan Watts, "Water Shortages Could Affect 5bn People by 2050, UN Report Warns," *Guardian*, https://www.theguardian.com/environment/2018/mar/19/water-shortages-could-affect-5bn-people-by-2050-un-report-warns.

226 **at seven hundred million:** "United Nations Decade 2010–2020," United Nations, https://www.un.org/en/events/desertification_decade/whynow.shtml.

226 **There's not much room for more:** Ibid.

226 **as their northern neighbors take their shares first:** Joel Bourne, Jr., "California's Pipe Dream," *National Geographic*, https://www.nationalgeographic.com/magazine/2010/04/plumbing-california/.

227 **will result in death:** "Can Humans Drink Seawater?," NOAA National Ocean Service, https://oceanservice.noaa.gov/facts/drinksw.html.

227 **it involves chasing drops of humidity:** Johannes Haarhoff, "The Distillation of Seawater on Ships in the 17th and 18th Centuries," *Heat Transfer Engineering* 30, no. 3 (2009), doi.org/10.1080/01457630701266413.

227 **They began investigating wastewater recycling:** David Gorn, "Desalination's Future in California Is Clouded in Cost and Controversy," KQED, https://www.kqed.org/science/1115545/desalination-why-tapping-sea-water-has-slowed-to-a-trickle-in-california.

228 **was launched in the mid-1970s:** Orange County Water District, *A History of Orange County Water District*, 2nd ed. (Orange County, CA: The Acorn Group, 2014), https://www.ocwd.com/media/1606/a-history-of-orange-county-water-district.pdf.

230 **as a by-product of industrial chemical processes:** "Technical Fact Sheet—NDMA," US Environmental Protection Agency, https://www.epa.gov/sites/production/files/2017-10/documents/ndma_fact_sheet_update_9-15-17_508.pdf.

231 **arsenic in the public water supply:** Sam Loewenberg, "The Poisoning of Bangladesh: How Arsenic Is Ravaging the Nation," The Undark, https://undark.org/article/bangladesh-arsenic-poisoning-drinking-water/.

231 **drink water polluted by pesticides:** "Millions Living in Rural France 'Are Drinking Polluted Tap Water,'" The Local FR, https://www.thelocal.fr/20170127/millions-in-rural-france-drink-polluted-tap-water.

231 **"lead to complaints":** Nives Dolsak and Aseem Prakash, "It's Not Just Flint: Here's Why We Ignore Water Pollution," *Washington Post*, https://www.washingtonpost

.com/news/monkey-cage/wp/2016/06/08/flints-contamination-and-victorias-secrets-heres-why-we-ignore-water-pollution/?noredirect=on&utm_term=.690f2f6a6b01.

231 **among other ways to spread the word:** What 2 Flush, http://www.what2flush.com/what-2-flush.html.

232 **is in need of around the globe:** William Sarni, *Water Tech: A Guide to Investment, Innovation and Business Opportunities in the Water Sector* (London: Taylor & Francis, 2013).

233 **protests wastewater reuse projects:** Sara Dolnicar, "When Public Opposition Defeats Alternative Water Projects—The Case of Toowoomba, Australia," *Water Research* 44, no. 1 (2010): 287–97, https://ro.uow.edu.au/cgi/viewcontent.cgi?referer=&httpsredir=1&article=1752&context=commpapers.

233 **the equivalent of twenty regular-sized water bottles:** Zero Mass Water, https://www.zeromasswater.com.

235 **the droplets trickle down and are stored in tanks:** Renee Cho, "The Fog Collectors: Harvesting Water from Thin Air," State of the Planet, Earth Institute, Columbia University, https://blogs.ei.columbia.edu/2011/03/07/the-fog-collectors-harvesting-water-from-thin-air/.

235 **with the extra water they capture:** Pat Evans, "This Chilean Brewery Makes Beer from Desert Fog," The Manual, https://www.themanual.com/food-and-drink/chilean-fog-beer-cerverceria-atrapaniebla//.

235 **for troops stationed in arid zones:** National Research Council et al., *Force Multiplying Technologies for Logistics Support to Military Operations* (Washington, DC: National Academies Press, 2014), chap. 3.

236 **mesh patterns found on fog nets:** Weiwei Shi et al., "Fog Harvesting with Harps," *ACS Applied Material Sciences and Interfaces* 10, no. 14 (2018): 11979–86, https://pubs.acs.org/doi/10.1021/acsami.7b17488; Ben Coxworth, "Move Over, Fog Nets—Fog Harp Is Better at Drawing Water from Air," New Atlas, https://newatlas.com/fog-harp/54031/.

Chapter 15: The Artificial Intelligence of Self-Driving Cars

239 **named Los Angeles as the most congested:** "Traffic Hotspot Study 2017," INRIX, http://www2.inrix.com/us-traffic-hotspot-study-2017.

240 **less concentrated areas of pollution:** Alissa Walker, "The Olympics Fixed LA's Traffic Problem—Can the 2028 Games Do It Permanently?," Curbed Los Angeles, https://la.curbed.com/2018/6/7/17419270/olympics-2028-los-angeles-1984-traffic.

240 **are exacerbated by the heated conditions:** Emily Guerin, "Take a Deep Breath and Read About How Bad LA Smog Really Is," LAist, https://laist.com/2018/10/03/take_a_deep_breath_and_read_about_how_bad_la_smog_really_is.php.

241 **than any other state:** "Auto Retailing: State by State," National Automobile Dealers Association, https://www.nada.org/statedata/.

241 **use the L.A. Metro system:** Vas Panagiotopolous, "The Los Angeles Metro Is Great—So Why Aren't People Using It?," City Metric, https://www.citymetric.com/transport/los-angeles-metro-great-so-why-aren-t-people-using-it-2742.

241 **are expected to reside in the city:** Emily Castor Warren, "A New Vision for Los Angeles Streets," Medium, https://medium.com/sharing-the-ride-with-lyft/a-new-vision-for-los-angeles-streets-74613e2f0dba.

241 **less than sixty hours in traffic:** "Los Angeles Traffic Is the Worst in the United States," INRIX, http://inrix.com/blog/2014/03/los-angeles-traffic-is-the-worst-in-the-united-states/.

241 **than a Los Angelean does today:** Ernst & Young Global Automotive Center, "Urban Traffic in 2050," Auto News, https://www.autonews.com/assets/jpg/anecongress/presentations/presentations/ANE-Congress-June-2012-Peter-Fuss.pdf.

241 **per driver, in the US alone:** "Traffic Hotspot Study 2017," INRIX.

242 **second-biggest cause of global warming:** Hannah Ritchie and Max Roser, "CO_2 and Other Greenhouse Gas Emissions," Our World in Data, https://ourworldindata.org/co2-and-other-greenhouse-gas-emissions.

242 **used to transport military supplies:** Encyclopaedia Britannica, s.v. "Roman road system," https://www.britannica.com/technology/Roman-road-system.

242 **most dangerous city in Europe for traffic accidents:** Jessica Phelan, "Rome Among Worst Cities in Europe for Road Safety, Traffic and Pollution: Greenpeace," The Local IT, https://www.thelocal.it/20180522/italy-rome-road-safety-traffic-air-quality-greenpeace.

242 **rarely taken into account by urban planners:** Angie Schmitt, "The Science Is Clear: More Highways Equals More Traffic. Why Are DOTs Still Ignoring it?," *Streetsblog* USA, https://usa.streetsblog.org/2017/06/21/the-science-is-clear-more-highways-equals-more-traffic-why-are-dots-still-ignoring-it/.

243 **about autonomous, or self-driving, vehicles:** Drive.ai, https://www.drive.ai/.

243 **and let's admit it, tempers tamed:** Jamie Condliffe, "A Single Autonomous Car Has a Huge Impact on Alleviating Traffic," *MIT Technology Review*, https://www.technologyreview.com/s/607841/a-single-autonomous-car-has-a-huge-impact-on-alleviating-traffic/.

243 **either texting or talking per hour of driving:** Aarian Marshall, "Turns Out, a Horrifying Number of People Use Their Phones While Driving," *Wired*, https://www.wired.com/2017/04/turns-horrifying-number-people-use-phones-driving/.

245 **it can begin to teach itself:** Nick Heath, "What Is AI? Everything You Need to Know About Artificial Intelligence," ZD Net, https://www.zdnet.com/article/what-is-ai-everything-you-need-to-know-about-artificial-intelligence/.

246 **convenient pickup and return points for online customers:** Jeanette Neumann, "Zara Turns to Robots as Its In-Store Pickups Surge," *Wall Street Journal*, https://www.wsj.com/articles/zara-turns-to-robots-as-in-store-pickups-surge-1520254800?itx%5Bidio%5D=5772297&ito=792&itq=b6ee309b-bb2b-4d79-96bc-1c1fbd87d4c4.

247 **takes control of anything away from them:** Lisa Fritscher, "Technophobia Is a Fear Related to the Loss of Control," Verywell Mind, https://www.verywellmind.com/what-is-the-fear-of-technology-2671897.

247 **have been involved in numerous crashes:** Ben Miller, "First Driverless Car Crash Study: Autonomous Vehicles Crash More, but Injuries Are Less Serious," Government Technology, https://www.govtech.com/fs/First-Driverless-Car-Crash-Study-Autonomous-Vehicles-Crash-More-Injuries-Are-Less-Serious.html.

247 **computers have a lot of room to do a better job:** "Road Safety Facts," Association for Safe International Road Travel, https://www.asirt.org/safe-travel/road-safety-facts/; World Health Organization, *Global Status Report of Road Safety 2018* (Geneva, Switzerland: WHO, 2018), https://apps.who.int/iris/bitstream/handle/10665/277370/WHO-NMH-NVI-18.20-eng.pdf?ua=1.

249 **twenty-one million miles of roads on Earth:** Asif Faiz, *Transportation Research Circular E-C167: The Promise of Rural Roads* (Washington, DC: Transportation Research Board of the National Academies, September 2012), http://onlinepubs.trb.org/onlinepubs/circulars/ec167.pdf.

250 **society is in for an abrupt confrontation:** "Employment by Major Industry Sector," US Bureau of Labor Statistics, https://www.bls.gov/emp/tables/employment-by-major-industry-sector.htm.

250 **and harm the state's economy:** Matt McFarland, "The Backlash Against Self-Driving Cars Officially Begins," CNN.com "Business," https://money.cnn.com/2017/01/10/technology/new-york-self-driving-cars-ridesharing/index.html.

251 **have faded into nonexistence:** "Aleppo, the Disappearing Memory of the Silk Roads," UNESCO Silk Roads Dialogue, Diversity & Development, https://en.unesco.org/silkroad/content/aleppo-disappearing-memory-silk-roads.

251 **Some versions will have us go even faster:** "Facts & Frequently Asked Questions," Virgin Hyperloop One, https://hyperloop-one.com/facts-frequently-asked-questions.

252 **or riff off it:** SpaceX and Tesla Motors, "Hyperloop Alpha," SpaceX, https://www.spacex.com/sites/spacex/files/hyperloop_alpha-20130812.pdf.

252 **have to be dug between end points to lay the tube:** Andrew Hawkins, "World's Third Hyperloop Test Track Is Now Under Construction," The Verge, https://www.theverge.com/2018/4/15/17235262/hyperloop-transportation-technologies-test-track-france.

Chapter 16: The City of the Future

255 **chronic heat waves of Saharan intensity:** Alessandro Dosio et al., "Extreme Heatwaves Under 1.5°C and 2°C Global Warming," *Environmental Research Letters* 13, no. 5 (2018), https://iopscience.iop.org/article/10.1088/1748-9326/aab827.

255 **forests will begin to grow in the Arctic:** David Spratt, "What Would 3 Degrees Mean?," Climate Code Red, http://www.climatecodered.org/2010/09/what-would-3-degrees-mean.html.

255 **the 1.5°C/2.7°F rise is assured:** Intergovernmental Panel on Climate Change, *Special Report: Global Warming of 1.5°C* (IPCC, 2018), https://www.ipcc.ch/sr15/.

256 **Green parks abound:** Masdar City, https://masdar.ae/en/masdar-city.

256 **in Abu Dhabi and around the world:** Masdar, https://masdar.ae/en.

256 **if the world weans itself off fossil fuels:** Yochi Dreazen, "UAE: Powering Down," Pulitzer Center, https://pulitzercenter.org/reporting/uae-powering-down.

257 **"The Gherkin" (because it looks like a cucumber):** "Norman Foster," Foster + Partners, https://www.fosterandpartners.com/studio/people/partnership-board/norman-foster/.

259 **less carbon dioxide emissions:** "Climotion," Oairo, https://www.oairo.me/bosch-climotion-hvac.

260 **to make the most out of the patches:** Abbas Hassan, "From Medieval Cairo to Modern Masdar City: Lessons Learned Through a Comparative," Research Gate, https://www.researchgate.net/figure/Masdar-City-consists-of-two-squares-The-first-is-big-and-the-second-is-small-The_fig3_276703027.

262 **and how they survive in extreme conditions—over centuries:** Glenn Dolcemascolo, "Cultural Heritage and Indigenous Knowledge," in *2017 Global Platform for Disaster Risk Reduction: From Commitment to Action, Cancun, Mexico, May 22–26* (UNISDR, 2017), https://www.unisdr.org/files/globalplatform/5921c5238b811 Cultural_heritage_and_Indigenous_Knowledge_WS_GP_2017_-_25_April.pdf.

262 **are high-tide lines and flood zones:** Lisa Hiwasaki et al., "Process for Integrating Local and Indigenous Knowledge with Science for Hydro-Meteorological Disaster Risk Reduction and Climate Change Adaption in Coastal and Small Island Communities," *International Journal of Disaster Risk Reduction* 10, no. A (2014): 15–27, https://doi.org/10.1016/j.ijdrr.2014.07.007.

264 **and washing machines. Black water is sewage:** Robert Lamb, "How Gray Water Reclamation Works," How Stuff Works, https://science.howstuffworks.com /environmental/green-science/gray-water-reclamation1.htm.

266 **into every half mile of land area:** "Hong Kong—the Facts," GovHK, https://www.gov.hk/en/about/abouthk/facts.htm.

266 **as opposed to cradle to grave:** "Design Approach," William McDonough & Partners, http://www.mcdonoughpartners.com/design-approach/.

267 **what we now know as urban environments:** Nancy Scola, "Google Is Building a City of the Future in Toronto. Would Anyone Want to Live There?," *Politico Magazine*, https://www.politico.com/magazine/story/2018/06/29/google-city-technology -toronto-canada-218841.

267 **the project's official description found—where else?—online:** Sidewalk Toronto, https://sidewalktoronto.ca/.

269 **"purpose-built for a new way of living":** NEOM, https://www.neom.com/.

270 **to caring for the sick and needy:** Cheryl Chumley, "Saudi Arabia's New City, Neom, a Mecca for Robots," *Washington Times*, https://www.washingtontimes.com/news /2017/oct/24/saudi-arabias-new-city-neom-mecca-robots/.

270 **around green spaces, bridges, and waterways:** Niall Walsh, "Insight into Secretive Unbuilt NEOM Megacity Ahead of Saudi Royal Visit," Arch Daily, https://www.archdaily.com/899430/insight-into-secretive-unbuilt-neom-megacity -ahead-of-saudi-royal-visit.

Conclusion

271 **Global carbon emissions in 2018 hit an all-time high:** "Carbon Budget 2018: An Annual Update of the Global Carbon Budget and Trends," Global Carbon Project, https://www.globalcarbonproject.org/carbonbudget/.

272 **They feared injury or death:** "History of Anti-Vaccination Movements," History of Vaccines, https://www.historyofvaccines.org/content/articles/history-anti -vaccination-movements.

273 **oppose the oversight initiative:** Jonathan Watts, "US and Saudi Arabia Blocking Regulation of Geoengineering, Sources Say," *Guardian*, https://www.theguardian .com/environment/2019/mar/18/us-and-saudi-arabia-blocking-regulation-of-geo engineering-sources-say.

Index

Note: Page numbers in *italic* indicate photographs and illustrations.

About the Author

Thomas M. Kostigen is a bestselling and award–winning author of numerous books, articles, and essays on the environment and social issues. He lives in Los Angeles.